LEED® GREEN ASSOCIATE™

V5 EXAM PREP

COMPLETE STUDY GUIDE

FOURTH EDITION
WITH EXAM SIMULATOR

A. Togay Koralturk

Projectific

Published by:
Projectific, Inc.
8 The Green
Suite #4445
Dover, DE 19901
www.projectific.com

ISBN: 978-1-972631-05-8 (paperback)

ISBN: 978-1-972631-06-5 (hardcover)

Library of Congress Control Number: 2026943737

Printed in the United States of America.

ALSO BY A. TOGAY KORALTURK

PASSING SCORE GUARANTEE

We are so confident that this study guide will prepare you for the LEED Green Associate V5 exam that we stand behind it with a money-back guarantee. If you study the material in this book, take the final full-length practice exam on the Exam Simulator, achieve a a score of 80% or higher (80 out of 100 correct answers) passing score on your first try, and still do not pass the official LEED Green Associate exam, we will refund the full purchase price of this book. Why first attempt only? Taking the practice exam more than once allows you to recognize questions and recall answers from memory rather than from understanding the material. Your first attempt is the only score that truly reflects whether you have learned the content in this guide.

Every question on the full-length practice exam in the Exam Simulator is drawn directly from the content covered in this book. The questions are designed to accurately reflect the difficulty, format, and scope of the actual LEED Green Associate exam. They are neither inflated in difficulty nor simplified.

Before starting the full-length practice exam on the Exam Simulator, you will be required to enter your full legal name. This name must exactly match the name that appears on your official LEED Green Associate exam score report.

Additionally, all of the following conditions must be met:

- You must purchase a new copy of this book directly from an authorized retailer. Used or borrowed copies are not eligible.
- You must sit for the official LEED Green Associate exam within fourteen (14) calendar days of completing the full-length practice exam on the Exam Simulator.
- This guarantee is available exclusively to customers residing in the United States of America due to banking and payment processing restrictions.

If you meet all of the eligibility requirements listed above, you can follow these steps to request your refund:

Submit your refund request via email to customercare@projectific.com within thirty (30) calendar days of receiving your official LEED Green Associate exam results. Include a copy of your official LEED Green Associate exam score report showing a failing status, with your full name and exam date clearly visible and your proof of purchase (receipt or order confirmation with an order number) for a new copy of this book. The publisher reserves the right to verify all submitted documentation. Upon verification of all submitted documentation, your refund will be processed within five (10) business days. Refunds will be issued in U.S. dollars to a U.S.-based bank account, PayPal, or via the original payment method where applicable.

This guarantee applies to one (1) refund per customer per book purchased. Multiple claims per individual are not permitted. The refund amount is limited to the retail purchase price of the book and does not include shipping, handling, taxes, the cost of the LEED Green Associate exam registration, or any other fees or expenses. The publisher reserves the right to verify all submitted documentation.

This guarantee is non-transferable and applies only to the original purchaser of the book.

The publisher reserves the right to modify or discontinue this guarantee program at any time. This guarantee provides a refund if you meet the eligibility requirements and do not pass the official LEED Green Associate V5 exam. It is not a promise or warranty that every reader will pass the LEED Green Associate exam, as exam performance depends on individual factors such as exam-day conditions. This guarantee is void where prohibited by law.

CONTENTS

PREFACE

The built environment shapes nearly every aspect of human life. What we build determines not only how resources are consumed, but also how we experience health, comfort, productivity, and quality of life. For that reason, green building is not merely a professional specialty or an academic topic. It is one of the most important ways to improve how people live and how the world's resources are used.

LEED is the most widely recognized framework for putting these principles into practice. My primary goal in writing this book was to create a single source that covers the complete LEED Green Associate exam scope and provides deep insight into the principles of green building, helping the reader to learn, rather than memorize, the essential information for the exam. This approach equips candidates to successfully pass the LEED Green Associate exam on the first try in order to actively apply green building and LEED principles in real life.

This book is written for a wide range of readers — from architecture and engineering students encountering LEED for the first time, to working professionals seeking formal credentials, to anyone with a genuine interest in sustainable design and construction. No prior knowledge of LEED or green building is assumed. Every concept is developed from the ground up.

Even though the LEED Green Associate exam is sometimes described as a conceptual exam, in reality the exam goes well beyond that. Without diving into the LEED prerequisites and credits, a candidate can never guarantee passing the test. For this reason, the book covers the core principles of each LEED credit category and then examines how those principles are applied through actual LEED requirements — bridging the gap between theory and practice.

As LEED v5 is quite different from previous versions, this book was entirely rewritten rather than updated from a previous edition. All chapters contain guidance on the aspects that require extensive attention, while pointing out finer details to focus on for exam purposes. Because this book teaches the real reasoning behind green building and LEED principles, the knowledge you gain will stay with you long after the exam.

To speed up your exam preparation, this book offers a streamlined system to minimize review time and significantly shorten your overall study. Once you finish reading, the *Key Things to Remember* sections will enable you to review the key points for the exam in no time. Additionally, all key terms and concepts appear in bold so you can easily find and review them. At that point, you can take the full-length practice exam before sitting for your actual exam.

My sincere hope is that the knowledge you gain here will serve you well throughout your career — not only on exam day.

Chapter 1:

Introduction to Green Buildings and Sustainability

According to the U.S. Environmental Protection Agency (EPA), green building is the practice of creating structures and using processes that are environmentally responsible and resource-efficient throughout a building's life cycle, from siting to design, construction, operation, maintenance, renovation, and deconstruction. This practice expands and complements the classical building design concerns of economy, utility, durability, and comfort. Green building is also known as a sustainable or high-performance building.

Green buildings are more environmentally sensitive, provide more comfort to building occupants, and also remain resource-efficient and high-performance buildings throughout their entire life cycle, but how does that happen? The answer will be explained throughout this book; however, to highlight the major concept of green buildings, we must first observe the natural system.

In the natural cycle, there isn't any waste. Even what we term "animal waste" is actually a sustainable product, which is an organic fertilizer in agriculture that improves plant nutrition and assists the growth of plants. The key factor of animal waste is its sustainability; it is produced by nature and is continuously reused in a never-ending loop. Nevertheless, is the waste of man-made products also like that? Think about a plastic bag, which takes about five hundred years to break down in the environment and creates a great deal of environmental harm during its lifetime.

Now, let us think about the "**built environment**", which refers to all the man-made surroundings that are needed for human activity, from roads, to buildings, to neighborhoods. We use most of the world's resources to develop our built environments, but do they end up like the plastic bag in the previous example, or can they continue sustainably in nature?

Green building is really a system that evolves by continuously enhancing its level of performance in order to get as close as possible to the natural system, which is by far the most efficient system known to man.

ENVIRONMENTAL IMPACTS OF BUILT ENVIRONMENTS

There is not a single building in the world that does not have an impact on the environment. That being said, the way we build can make a great contribution toward reducing this impact.

In the Unites States, buildings account for the following:

- 14% of potable water consumption[1]
- 30% of waste output
- 40% of raw materials use[2]
- 38% of carbon dioxide (CO_2) emissions
- 24% to 50% of energy use
- 72% of electricity consumption[3]

Accounting for 38% of carbon dioxide emissions, buildings are on top of the list and produce even more carbon dioxide (CO_2) than transportation and industrial emitters. And all the built environment contributes to 67% of all greenhouse gas emissions.[4]

But why do our buildings harm the environment? First, when we clear the land to make way for a development, the wildlife habitat in that area is destroyed, and the preexisting permeable soil becomes an impermeable surface. Second, we manufacture building products that pollute water and air, release toxic chemicals, and emit greenhouse gases. Furthermore, during the building operations phase, the building consumes energy, which further increases greenhouse gas emissions and adds to the consumption of potable water, which later becomes a waste product. With the commuting of building users and visitors to the building, the use of single-occupancy vehicles results in increased energy consumption and greenhouse gas emissions.

The list of aspects related to environmental impact is seemingly endless.

According to the United Nations, **sustainable development** means meeting the needs of the present without compromising the ability of future generations to meet their own needs.[5] The **United Nations Sustainable Development Goals** (**SDGs**) aim to end poverty, protect the planet, and ensure peace and prosperity for all by 2030.

The following are the 17 SDGs:

The **three pillars of sustainability**—economic prosperity, social equity and justice, and environmental stewardship—are embedded in these 17 goals. Green building plays a role in achieving virtually all of them.

BENEFITS OF GREEN BUILDINGS

When looking at the statistics of green buildings, a study by the New Buildings Institute found that average energy use intensity of green buildings is 24% lower than typical buildings.[6]

Additionally, a U.S. General Services Administration survey among twelve green buildings states that green buildings have:

- 26% energy use reduction
- 27% higher levels of occupant satisfaction
- 13% lower maintenance costs
- 33% lower carbon dioxide emissions

Occupants of green buildings are exposed to far lower levels of indoor pollutants and are exposed to superior air, lighting, and acoustical qualities than occupants of regular buildings.

According to the U.S. Environmental Protection Agency (EPA), people in the United States spend, on average, 90% of their time indoors.[7] Which means that developing green buildings will make a serious contribution to the health and comfort of the building occupants and their productivity.

REGENERATIVE DESIGN

Regenerative design goes beyond sustainable design. While sustainable design aims to minimize harm and reduce the negative impacts of buildings on the environment, regenerative design aims to create a net positive impact—giving back more than it takes. **Regenerative (restorative) design** is an approach to building in which the project is designed to produce more energy than it consumes, generate no waste, and actively restore the ecosystems and habitats that development has degraded. Excess energy produced on site can be returned to the utility grid, making the building a contributor to the energy supply rather than just a consumer. Projects that produce at least as much renewable energy as they consume on an annual basis are often referred to as **net-zero energy buildings**. Regenerative design takes this further by also addressing water, materials, and ecological health—restoring natural systems, improving biodiversity, and leaving the surrounding environment in a better condition than before construction began.

TRIPLE BOTTOM LINE

All green buildings should establish three components:

- Economic prosperity (profit)
- Social equity and justice (people)
- Environmental stewardship (planet)

Those three components are called the **triple bottom line**, also referred to as the **three pillars of sustainability**. A commitment to the triple bottom line means looking beyond a project's profitability, or considering more than just a project's sustainable features, or social responsibility.

When developing a green building, the main goal is to establish all three components at the same time. A green building should have an increased value, which will result in increased profit, an environmentally friendly design, and should also serve its users and the community well.

To illustrate the concept, let's assess the triple bottom line of a skylight that will be installed on the roof of an office space. A skylight will reduce the energy consumed by lighting fixtures, which will reduce greenhouse gas emissions. Less energy usage will also mean fewer operating costs for the office building. Also, natural daylight will create a healthier office space for workers.

If we are able to address and achieve all three components of the triple bottom line, why not install the skylight?

GREENHOUSE GAS EMISSIONS, CLIMATE CHANGE AND GREEN BUILDINGS

The primary greenhouse gases in the earth's atmosphere are water vapor, carbon dioxide, nitrous oxide, methane, and ozone. Each of them is essential to absorb and emit radiation, and without them, the average temperature of the earth's surface would be much colder than it is now.

However, with the increasing concentration of carbon dioxide in the atmosphere, which is the result of human activity, the greenhouse gas balance has been disturbed. This increase has become the number-one cause for global climate change. As the greenhouse gas emissions increase, an even larger buildup of carbon dioxide occurs in the atmosphere, thus warming the climate further.

As discussed, the built environment contributes to 67% of all greenhouse gas emissions.[4] This comes from building systems and energy use, transportation, water use, changes in the land cover, building products, and construction. Although changing the way we build will play a significant role in reducing greenhouse gas emissions, merely developing green buildings will not be enough. To establish a lower carbon emissions rate, we need to develop green communities as well.

The first thing that a green community should implement is the reduction of the distances that vehicles travel. The buildings in the suburban areas, even if they are green buildings, create greenhouse gas emissions indirectly due to the long distances that people drive every day in order to commute to and from work. Additionally, emissions from transportation make up half of the total emissions associated with the project, which also means that the project location is another critical factor to consider when developing a green building.

Table 1 shows the common sources of greenhouse gas emissions according to their category, which is important to know for exam purposes.

Common Sources of Federal Greenhouse Gas Emissions	
Category	**Sources**
Scope 1 Energy	Vehicles and equipment
	Stationary sources
	On-site landfills and wastewater treatment
	Fugitive emissions
Scope 2 Energy	Purchased electricity
	Purchased heating/cooling
	Purchased steam
Scope 3 Energy	Transmission and distribution losses from purchased electricity
	Business travel
	Employee commuting
	Contracted solid waste disposal
	Contracted wastewater treatment

Scope 1 energy relates to the direct energy from the owned or controlled sources, **scope 2 energy** relates to the purchased energy, and **scope 3 energy** relates to energy sources that are not owned or directly controlled. To illustrate, the energy generated on-site through burning of fossil fuels falls under scope 1 energy, and the resulting greenhouse gas emissions are classified as scope 1 emissions. The electricity bought from a utility company is classified as scope 2 energy, and the resulting greenhouse gas emissions are scope 2 emissions.

SUSTAINABLE THINKING

The green building process is very different from the conventional building process. While the conventional building process focuses on the snapshots of the project, the green building process looks at the project as a whole. There are three major concepts integral to both green buildings and sustainability that are not a part of the conventional building process: systems thinking, life-cycle thinking, and integrated process.

As all the buildings are composed of different interrelated systems that form a whole, each of them has an impact on the others. Consider a building envelope and an HVAC (heating, ventilating, and air conditioning) system. The insulation level that a building envelope provides will affect the size and operating cost of the HVAC system. The better the insulation, the smaller the HVAC system will be, which will also mean lower operating costs. Meanwhile, a proper waterproofing membrane on the building foundation will protect the foundation and affect the durability of the whole building. This will also result in less humidity inside the building, which will affect both the HVAC system and the health of the building occupants.

The systems of a building include resources, materials, energy, building occupants, and information. The term **systems thinking** refers to the understanding of each and every system of a building while also understanding their relationships and looking at the project as a whole.

The second major concept of sustainable thinking concerns the **life-cycle approach**, which evaluates the entire life of a project, product, or service. For example, when selecting a building product, the project teams need to ask the following questions: What is the raw material of the product? Was it extracted in a responsible manner? How long can the product be used? How much maintenance does it require? What will happen to the product once its useful life ends? Can it be recycled?

Integrated process, which is the third major concept of sustainable thinking, emphasizes the importance of connection and communication among all the professionals and stakeholders in the project. To be able to see the big picture, all the different people in the project should act as a single team. The stakeholders, architects, structural engineers, electrical engineers, mechanical engineers, facility managers, etc.—should work collaboratively, starting from the predesign phase of the project. That way, each team member can understand the perspectives and strategies of the rest of the team. And the collaboration will enable better decisions to be made to create a high-performance building. Let's take a closer look at these three major green building concepts.

SYSTEMS THINKING

The main idea behind systems thinking is to see that the systems in the building do not work in isolation. Actually, there is not a single system in our universe that works in isolation; everything is interconnected. In order to effectively understand systems thinking, we need to understand the types of systems.

There are two types of systems. The first type is the **open system**, which is a system that constantly consumes other items, uses them, and produces waste at the end. A whole city may be an example of this, as it imports food, energy, and products; however, while a city uses products, it also produces waste, often in the form of sewage, pollution, and solid waste, which consequently mark the end of the system. Therefore, there are no **feedback loops** in open systems.

However, in nature, everything is in a loop, and thus no open system exists. Therefore, this is called a **closed system**, which does not produce any waste product at the end. The system continues independently. Think about the water cycle in our world: The sun heats and evaporates the water on the oceans and the seas and then creates water vapor. The resulting water vapor rises in the air, which condenses into rain and goes back to the earth's surface. That water then becomes underground water, which again joins the oceans and seas with the flow. This sustainable cycle does not end.

Finding ways to develop closed systems in our buildings will result in creating sustainable buildings. In order to establish that, we need to evaluate the individual elements in addition to their relationships to each other as a whole when designing our buildings. A decision made for an individual element will have a ripple effect that will affect other elements and systems. Remember the example about installing a proper waterproofing membrane on the building foundation which affects the foundation, the durability of the whole building, the humidity level inside the building, the HVAC system, and the health of building occupants.

Understanding the **emergent properties** of systems is also essential in order to better evaluate why we need to form systems. Every system is a composition of its individual parts. Let us take the human body as an example, in which cells form tissues, tissues form organs, organs form the organ systems, and organ systems form humans. A lung is made up of cells; however, having only lung cells together will not do anything. The whole system needs to be there. Certain properties emerge as the result of interaction of individual elements; these are called **emergent properties** of the systems, and only systems can have emergent properties, not the individuals. The breathing function of the lung is the emergent property of a whole successful system. This is why systems thinking requires the creation of emergent properties to achieve rewarding outcomes.

To take the discussion to the next level, we will now examine feedback loops, followed by leverage points.

FEEDBACK LOOPS

Feedback loops refer to the flow of information within a system that allows the system to take action. There are two types of feedback loops in systems: positive and negative.

A **positive feedback loop** can be summarized as A producing B, which in turn produces more of A. An example of this would be an interest-earning savings account. As the account grows, more interest is earned, which in turn brings further account growth.

In a **negative feedback loop**, a change brings an additional change in the opposite direction. If a room gets warmer than the set temperature, the thermostat will send a signal to the air conditioning, and the air conditioning will stop blowing warm air.

Understanding these feedback loops is crucial for designing sustainable buildings and neighborhoods. With the feedback loops providing the flow of information, the system can take action when there's a problem. Since the positive feedback loops don't stop, they may create problems in the systems if they're unchecked. However, if the positive feedback loops are checked with the negative feedback loops, the system can stop if there is too much of something.

With current technology, both feedback loops can be implemented by the use of sensors, controls, building automation systems, and more. A building automation system can show building management the excess energy usage statistics by tracking the individual consumption of all the building systems. Equipped with only this information, the building management can then respond and try to discover ways to reduce the energy use. This is called the **Prius effect**, which means that users can respond to something only if they have real-time information about it.

LEVERAGE POINTS

A small change can yield big changes. In other words, any action taken in a **leverage point** of the system can bring about significant results in the system. Think about a construction site that is located close to a river that is the main irrigation source of the surrounding farmlands. If the construction site does not take any action to stop sedimentation and pollution resulting from the construction activities, the whole river can become contaminated. This will then contaminate the farmlands, which will result in contaminated foods. However, by taking action to avoid sedimentation and pollution to the river, which is really a small action when considering the outcome, a whole river can be saved that will also result in saving the farmlands and human health.

Being able to see where actions can yield big changes is a very important part of systems thinking, and only by asking the right questions can project teams gain the insight needed to move toward sustainability.

LIFE-CYCLE APPROACH

The second major concept of sustainable thinking is the life-cycle approach. Green buildings need a life-cycle approach, and project teams should always look at the big picture by evaluating all phases of the project together rather than merely looking at the snapshots. A life cycle of a building covers location selection, design, construction, operations and maintenance, refurbishment, and renovation or demolition. At the very beginning of the project, project teams need to think about the very last phase of the building, which would be the demolition or complete renovation.

Let's consider a wood-flooring product that is selected to be used in the project. Project teams need to evaluate the raw material manufacturer of the wood flooring. Was the wood extracted in a responsible manner? After cutting trees, does the wood manufacturer plant new trees in order to avoid deforestation?

Then, the project teams also need to consider the cost and durability of the product, and they need to evaluate what can happen to the wood flooring after its useful life. Can it be recycled, or will it become a waste product? This practice, which investigates materials from their *extraction* to their *disposal*, is called a **cradle-to-grave approach.**

In order to avoid waste, closed systems should be formed, and a product should become a part of another product, after its useful life, through recycling or reusing. Evaluating products according to this philosophy is called a **cradle-to-cradle** approach, which aims to extend the product lifecycle to avoid waste.

To examine the environmental cost of a product to nature, the energy consumed resulting from a product's manufacturing, transportation, installation, and use should also be evaluated. The total energy consumed in all of these stages is called the **embodied energy**. By way of illustration, think about marble mined in China, manufactured in Europe, brought to the United States for purchase, and then bought for use in a building. The embodied energy of that marble will include all the energy consumed for the extraction, manufacture, transportation, installation at the site, and, finally, disposal. The more embodied energy a product contains, the more it will result in damage to the environment.

Closely related to embodied energy is **embodied carbon**, which measures the total greenhouse gas emissions associated with a product's entire life cycle—from raw material extraction and manufacturing to transportation, installation, and disposal. While embodied energy focuses on how much energy is consumed, embodied carbon focuses on how much carbon dioxide and other greenhouse gases are released in the process. Two products can have similar embodied energy but very different embodied carbon if one is manufactured using renewable energy and the other using fossil fuels.

As buildings become more energy efficient during operations, the share of a building's total carbon footprint that comes from its materials and construction becomes proportionally larger. This makes embodied carbon an increasingly important consideration. However, reducing embodied carbon cannot be treated in isolation from operational carbon. **Operational carbon** refers to the emissions produced by running the building on a daily basis, such as energy used for heating, cooling, lighting, and powering equipment. A highly insulating material, for example, may significantly reduce heating and cooling energy over the building's lifetime but carry a heavy carbon footprint from its manufacturing process. Finding the right balance requires project teams to evaluate both dimensions together, ideally through early and ongoing collaboration so that carbon impact can be optimized at every stage of design.

When deciding on the products to be used in projects and also to support decision-making of the project teams, the life-cycle approach should be implemented for environmental considerations with life-cycle assessment. And for cost considerations, life-cycle costing should be applied.

Life-cycle assessment (**LCA**) (also discussed under Life-cycle Assessment) evaluates all the environmental effects of a product during its whole lifetime. A cradle-to-grave, or cradle-to-cradle, approach is used in LCA, and the total energy use and other environmental consequences resulting from the creation of that material are calculated. Hence, conducting an LCA for the whole building is beneficial in showing the trade-offs between different materials and in helping in the selection of materials that would be the best fit for the project and the environment.

Life-cycle costing (**LCC**) assesses a product's total cost by evaluating its initial cost, service cost, maintenance cost, operating cost, and disposal cost. For example, a more expensive but more durable refrigeration system can cost less over time compared to a less expensive but less durable one. If the less durable refrigeration system will require more maintenance and will have a shorter lifetime, using that system may result in more total expenditure for the project owner.

Life-cycle thinking should be applied to all decisions made in green building, not only to product selection. Setting targets and discovering ways to find the best solutions early in the project can yield great results.

INTEGRATED PROCESS

The third and final major concept of sustainable thinking is the integrated process, which is the key to applying sustainable thinking to real-life situations.

Remember the example about the system of the human body, in which the lung is made of lots of cells, but when the cells are put together, without the whole system in place, the cells cannot perform the breathing function of the lung. Without a working system, an emergent property cannot be created.

Assembling the project owner, key stakeholders, architects, civil engineers, and mechanical and electrical engineers will not create a successful project; rather, a system should be formed in which there is a flow of information and collaboration between each member for each phase of the project. To better understand the topic, we will now discuss the conventional building process and then return to explain the details of the integrated building process.

CONVENTIONAL BUILDING PROCESS

In the conventional building process, project teams work in isolation and collaborate when problems occur. Usually, the process starts with a project owner hiring an architectural firm to develop the design. As the design progresses, the architectural firm hires electrical and mechanical designers, who then start integrating the electrical and mechanical parts of the design separately. Then construction plans are issued, and the owner starts the bidding process. In this phase, different general contractors are requested to bid, and once all the bids are submitted, the owner chooses the general contractor and awards the construction of the project. This is called the **design-bid-build** process.

With this approach, the general contractor was not involved in the design phase at all, and furthermore, the general contractor didn't have any collaboration with the architectural firm. In addition, the general contractor didn't have any collaboration with the mechanical and electrical designers. During the construction phase, as problems arise, the general contractor will invite the architectural firm to provide input. In turn, the architectural firm might invite the electrical and mechanical designers to the meetings in order to solve the problems at hand. The outcome of this process will usually be construction delays and extra project costs.

Referring to the human body system example, all the cells were there, but the system was not working.

INTEGRATED BUILDING PROCESS IN DETAILS

To establish an integrated building process, a working system needs to be created. And the system should be active as early as possible during the **predesign phase** or even before deciding on the project location. Think about the following questions:

- How is the climate in the project's location, and what type of heating and cooling system should be installed?
- How should the building be oriented in order to get more sunlight into the building and to reduce the operating costs?

- Is there enough rainfall at the location to capture and use rainwater for irrigation?
- What are the goals and requirements of the project's owner?
- How will the building occupants commute once the project is finished, and what are the public transportation options?
- How will the resources and construction materials be transported to the project site?
- What are the building code requirements at the project's location?
- How does energy get to the site?
- What is the type of soil on-site?
- What type of vegetation exists on-site?

They're really the type of questions that should be addressed during the earliest phase of the project. This is why the key in the integrated process is to integrate the whole project team early in order to develop an efficient project design and allow the project team to discover strategies. Taking actions late in the process will only result in delays and additional costs.

IMPORTANCE OF THE PROJECT'S LOCATION

It might be noticed that most of the questions from the previous topic depend on the project's location. To turn the project goals into reality, a strong evaluation of the project location is mandatory. The project location should assist the project teams rather than creating challenges.

The preferred sequence is to set the project goals first and then choose the location that best fits those goals. If the sequencing is vice versa, project teams should still need to discover ways to establish project goals; however, they can also face more problems.

Understanding the whole community is essential for successful project integration. If people need to drive single-occupancy vehicles to commute to a building, the project will generate more greenhouse emissions. The climate in the project location is another factor, which will impact all the building systems. The soil type will also change according to the location, which will affect the whole structural calculation of the building and lead to changes in the whole structural design.

Even the level of the underground water for a particular location can make a huge impact on the design and construction of the basement levels, as well as the whole building. When evaluating the site, it is always recommended to work with experts such as engineers, hydrologists, ecologists, and project development consultants. Evaluating zoning and local codes, International Code Council (ICC) standards, Americans with Disabilities Act (ADA) requirements, and all other regulations at that location in the beginning of the project is also essential. Most of the mentioned laws will depend on the project's location, and they will take precedence over LEED requirements. For example, zoning and local codes—not LEED—will determine the building height, type of project to be constructed, size of the building, and minimum parking requirements for a particular location.

MAIN PRINCIPLES OF THE INTEGRATED PROCESS

Below is a list of the main principles of the integrated process:

- **Process matters**: The system should work efficiently.
- **Start early**: The system should be active as early as possible.
- **Follow through**: The commitment should continue throughout the life of the project. It should not end with the completion of construction. This is to ensure that the project strategies are maintained and all the building systems are still efficient through the building operations phase.
- **Look beyond initial costs to long-term savings**: Green strategies have very short payback periods. And project stakeholders should be informed that their project would actually cost less in the long run.
- **Include and collaborate**: The professional project teams should work together with members of the community and always look at the big picture.

To establish a working system among the project team members, stakeholders and member of the community, an iterative process needs to be developed.

ITERATIVE PROCESS

In the integrated process, the ideas are developed by the entire project team. With the discovery of an idea, smaller groups are assigned to research and refine the topic, and then the idea is brought back to the table for the whole team to make the final decision. Hence, an iterative process exists that contains lots of feedback loops in order to establish a working system.

Following is a list of stages for the iterative process:

- Establish clear goals.
- Brainstorm and develop creative and effective solutions.
- Research and refine ideas.
- Explore synergies between different strategies.
- Establish metrics to measure success.
- Set new goals based on the work that has been done.

This approach is unlike the conventional building approach, in which a project team member completes a work individually and then passes it to the next person (which is called a **linear approach.**) To establish an integrative process, it is useful to conduct different types of meetings. Such meetings would include charrettes, team meetings, small task groups, and stakeholder meetings.

Charrettes are intense workshops that are generally held in the beginning of the project and during project milestones. These workshops bring together the entire project team, **facility manager**, and all the stakeholders and outside experts/consultants (the facility manager is the person who will be responsible for running the operations of the buildings after the completion of construction.) Charrettes allow brainstorming and collaboration among different disciplines and experiences. An agenda and list of project goals are discussed in charrettes in order to make important decisions and agree on project goals. With brainstorming and collaboration of the team members, charrettes will also allow the discovery of new project goals. Since charrettes are highly structured, a facilitator is also recommended.

Below are the typical stages of a charrette:

- Background briefing: To make sure all participants have the basic information about the project and the topics that will be discussed
- Brainstorming
- Synthesis of work: Discussion of concerns and development of recommendations
- Initial response from the owner or the developer about the discussed strategies and goals
- Follow-up: A written report stating the identified actions that should be sent to all participants

Team meetings will be helpful in discussing strategies among team members from different disciplines. As discussed, the design of a building envelope will have a great impact on the building's HVAC system. The level of insulation that a facade provides will affect the heating and cooling loads of the building, which will affect the size and type of the HVAC system to be used in the project. Thus, in a team meeting, the facade designer and MEP designer can work collaboratively to discover the best design solutions for both the facade and the HVAC system.

Small task groups can be held inside each discipline and, together with the involvement of consultants, if they are needed, the groups can explore particular topics, research, and refine ideas. Then, the results can be shared with the whole project team.

Stakeholder meetings are conducted among the project team, stakeholders, neighbors, and community members in order to understand and discuss community needs, issues, and concerns.

It is also essential to involve the facility manager in the building design. Since the facility manager will be responsible for operating the whole building after the completion of construction, facility manager's ideas can result in a better building design and a more efficient building operation phase.

As mentioned, in the conventional building process, a **design-bid-build** approach is used. The construction team is not even a part of the project team during the design phase, and, moreover, the predesign phase does not even exist. However, with integrated project delivery, all the stakeholders, design teams, and construction teams, who really make up the whole project team, work collaboratively from the beginning of the project. This approach is called **design-build**.

TEAM SELECTION

The outcome of the project will depend on the professionals working on the project and the system created. Team members with green building knowledge and experience will make a big impact on the results. LEED Green Associate and LEED AP BD+C credentials can be sought in the team members to ensure adequate knowledge.

The duty of a LEED AP BD+C project team member is to support and encourage integrated design and streamline the LEED application and certification process. A LEED Green Associate team member will have the general knowledge of green building practices and will provide support to other project team members. (It is very important to remember their duties for exam purposes.) If there are inexperienced people on the team, it will be necessary to provide training and orientation.

There will also be a LEED Project Administrator on the team who will be assigned during the LEED project registration and will be the primary contact of GBCI for the LEED project. The LEED Project Administrator is a team member who will act as a project manager, overseeing the LEED certification process. (The LEED Project Administrator will be discussed in more detail under Project Registration.) To avoid confusion, a LEED Project Administrator does not need to be a LEED Green Associate or a LEED AP BD+C. The duties of LEED Green Associates, LEED APs, and LEED Project Administrators in the project are completely different.

EVALUATION AND SELECTION OF STRATEGIES AND TECHNOLOGIES

When the goals are set for the project, there should be a system in place to evaluate the different alternatives. Tools such as life-cycle assessment, life-cycle cost analysis, and computer modeling can all provide great benefits.

Take **Building Information Modeling (BIM)** as an example, which allows the creation of a 3D project model. With the use of BIM, project teams can model their project in a 3D model, define the types of materials they plan to use, set the project location, and run energy simulations for the building according to the climate of the project location. Alternatively, the project teams can see how the sunlight interacts with the building and adjust the building orientation in order to maximize the benefits of sunlight. The location of the windows and skylights also can be evaluated from the 3D model, which can result in providing natural sunlight to the building occupants and a reduction in lighting costs. With the integration of the project schedule to the 3D model, 4D construction simulations can be developed, and the project teams can find more effective ways to reduce the total construction duration.

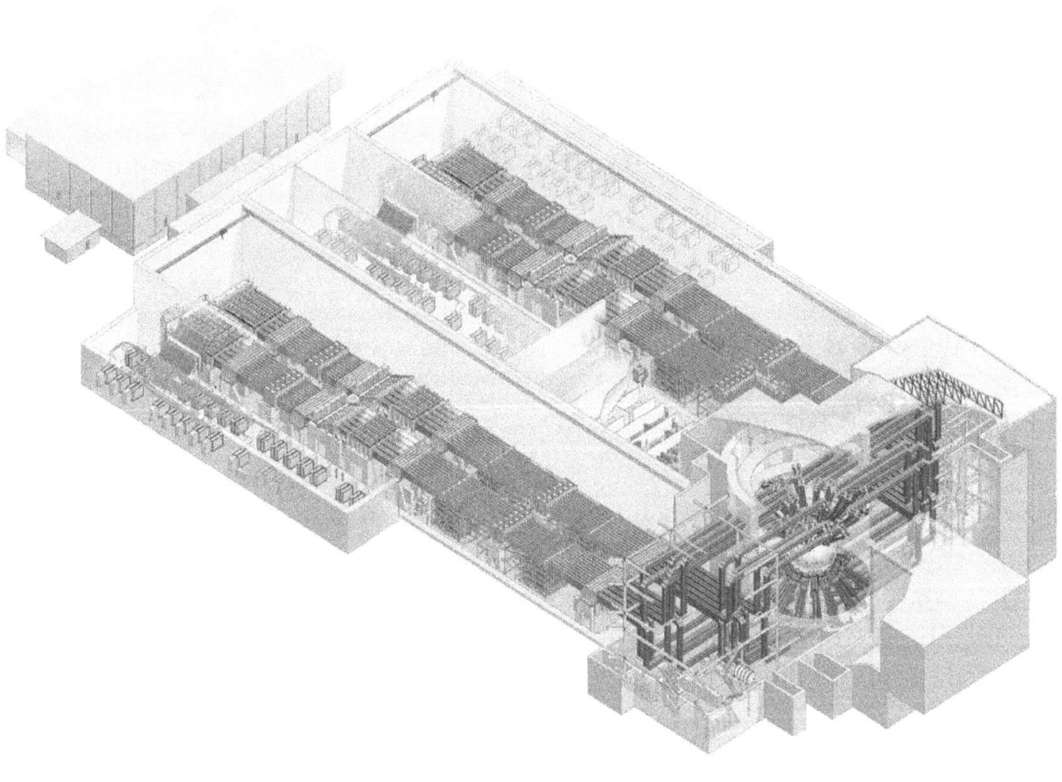

BIM model

When doing evaluations about project strategies and technologies, project teams should also refer to the triple bottom line, which is another important aspect for all green buildings.

The phases of the "evaluation" in the integrated process are summarized below:

- Set goals.
- Benchmark performance.
- Identify opportunities for improvement.
- Prioritize and align improvement opportunities with the project goals.
- Implement the program.
- Measure performance and use third-party verification.
- Set new goals or revise goals.

IMPLEMENTATION

Once the design and project planning phases are complete, it is time to turn the green building into a reality by stepping into the construction phase. All the work that was done to that point will make a great contribution to the construction phase as it will both reduce the duration of construction and the total money spent for the project. Documentation in this phase will be very important for compliance with regulatory requirements and achieving a LEED certification.

Let's take a look at some of the main factors that should not be overlooked during construction:

- Prevent air and water pollution by implementing an **erosion and sedimentation control** (**ESC**) plan that also includes a **stormwater pollution prevention plan** to prevent erosion, sedimentation, and stormwater pollution to the water bodies, wetlands, and entire neighborhood.
- Develop an **indoor air quality management plan** that will protect the indoor air quality for construction workers and building occupants. Typical strategies will include protecting the HVAC system from dust during the construction phase, using materials with minimal levels of toxicity, ventilating the whole building to clear construction contamination before occupancy, and more.
- Develop a **waste management plan** in order to address the storage, collection, and disposal of waste generated during the construction phase.

The costs during the construction phase will be classified either as **soft costs** or **hard costs**. Soft costs cover everything needed for developing a project that does not physically contribute to the building. All the management and supervision costs, design costs, permits, and taxes can be seen as soft costs, while all the other costs that physically contribute to the construction, such as labor costs, construction materials costs, and equipment costs, are hard costs.

Landfill *(Photo courtesy of Cezary)*

Even if construction is completed and both the design and construction teams leave, the integrative process should continue. And in order to continue the integrative process throughout the whole lifetime of a building, the project team needs to be involved in the training and orientation of the building operations personnel, and every single bit of information about the building should be passed on to them.

ONGOING OPERATIONS AND MAINTENANCE

This is the phase in which every project goal, all the assumptions that were made during the design, and the performance of all the building systems should be continuously verified in order to ensure high building performance. In order to establish a successful operations and maintenance (O&M) phase, a system needs to be created with the following activities creating a loop:

1. Audit
2. Plan
3. Measure
4. Adjust and go back to the audit

The LEED O+M certification, which is discussed under the Rating Systems Descriptions, is all about ensuring an efficient ongoing building performance. Projects need to get re-certification every three years in order not to lose their LEED O+M certificate.

The key to understanding whether the building is performing sustainably is to track and evaluate data that the building systems provide. If there's a building automation system in place, the real-time data it provides can show opportunities for energy and water savings and can point out a problem in a particular part of a building. Without such data, it would be impossible to track building operations. This is the reason why individual energy meters, individual water meters, a building automation system, and tracking devices are needed to successfully operate a building. Additionally, to reveal problems, the following types of inspections can be conducted:

- Retrocommissioning (this is a systematic process for analyzing and optimizing building system performance)
- Energy and water audits
- Solid waste audits
- Occupant surveys
- Green purchasing and green housekeeping program assessments

Orientation and training of the building personnel and occupants should be repeatedly conducted, as tenants and personnel change, and new lessons are learned. Educating building occupants about the green features of the building and sustainable practices can encourage them to participate in sustainable building operations. Continuous feedback from all the building users and personnel is needed to discover potentials for improvement.

The success of the O&M phase will depend on the following principles:

- Start early.
- Find the right team and create the right system.
- Understand the system across time and space.
- Develop measurable goals.
- Follow an iterative process to achieve goals.
- Commit to continuous improvement.

PHASES OF THE INTEGRATED PROCESS

The whole integrated process can be summarized by three main phases:

1. **Discovery phase**: This is the most important phase, which also includes predesign to ensure green building goals are set early and cost-effectively. The risks are identified, and the triple bottom line of the project is assessed. A system is created that enables the efficient collaboration of every participant in the project. The location of the project is well chosen, and applicable codes and standards are evaluated. The level of LEED certification desired and its requirements are thoroughly considered in this phase. Life-cycle costing, life-cycle assessment, and computer modeling of the project are used to evaluate different strategies and alternatives.
2. **Design and construction phase**: The system continues to work, and charrettes, team meetings, small task groups, and stakeholder meetings are continuously conducted. Necessary actions required to protect the construction workers, building occupants, neighborhood, and surrounding environment from construction pollution are taken. The green building goals are turned into reality, with the exception of maintaining them throughout the whole life cycle of the building. All the requirements for LEED certification are documented. The project team conducts both the training and orientation of the building operations personnel.

3. **Occupancy, operations, and performance feedback phase**: The performance of all the building systems should be checked throughout the lifetime of the building. The data that the building systems provide are monitored and evaluated. Orientation and training of the building personnel and occupants are conducted repeatedly as tenants and personnel change and new lessons are learned. Building occupants are educated about the green features of the building and sustainable practices. Continuous feedback from all the building users and personnel is received and the system continues for the whole life of the building.

END OF CHAPTER EXERCISES

EXERCISE: MATCHING GAME

Match the following terms to their definition.

Terms		
Regenerative Design	Triple Bottom Line	Life-Cycle Assessment (LCA)
Operational Carbon	Charettes	Hard costs
Embodied Carbon	Life-Cycle Costing (LCC)	Systems Thinking

1) ____________________________: A framework that evaluates a project's success across three dimensions: economic prosperity (profit), social equity and justice (people), and environmental stewardship (planet).

2) ____________________________: A method that estimates a product's total cost over its life, including initial cost, operation, maintenance, service/repairs, and end-of-life disposal.

3) ____________________________: Costs that directly pay for the physical construction of the building, including labor, construction materials, and equipment.

4) ____________________________: Intense workshops that are generally held at the beginning of the project and during project milestones. These workshops bring together the entire project team, facility manager, and all the stakeholders and outside experts/consultants.

5) ____________________________: A method that quantitatively evaluates a product's environmental impacts across its life cycle.

6) ____________________________: Refers to the understanding of each and every system of a building while also understanding their relationships and looking at the project as a whole.

7) ____________________________: Emissions produced by running the building on a daily basis, such as energy used for heating, cooling, lighting, and powering equipment.

8) ____________________________: A design approach where the project is intended to give back more than it takes—producing more energy than it consumes, generating little to no waste, and actively restoring ecosystems and habitats that development has degraded.

9) ____________________________: Measures the total greenhouse gas emissions associated with a product's entire life cycle—from raw material extraction and manufacturing to transportation, installation, and disposal.

ANSWER: MATCHING GAME

1) Triple Bottom Line	4) Charettes	7) Operational Carbon
2) Life-Cycle Costing (LCC)	5) Life-Cycle Assessment (LCA)	8) Regenerative Design
3) Hard costs	6) Systems Thinking	9) Embodied Carbon

PRACTICE EXAM

This book includes end-of-chapter practice exams and a full-length practice exam, all available through our online exam simulator to help you become familiar with the format of the actual LEED Green Associate exam. The end-of-chapter practice exams are designed to assess your understanding of each chapter and build your confidence with the exam format.

To access the end-of-chapter practice exams and the full-length practice exam, please refer to the *How to Access the Exam Simulator* section at the back of the book for instructions on creating your password. Once your password has been created, you can log in to the e-learning platform and take the practice exams by scanning the QR code below with your laptop, phone, tablet, or another compatible device. If you have trouble scanning the QR code, visit www.projectific.com/account and log in using your email address and password. If you need any assistance, please email us at customercare@projectific.com.

CHAPTER 2:

U.S. GREEN BUILDING COUNCIL AND ITS PROGRAMS

Founded in the early 1990s and based in Washington, DC, the U.S. Green Building Council (USGBC) is a nonprofit organization committed to a sustainable future through cost- and energy-efficient green buildings.

USGBC is made up of tens of thousands of member organizations, chapters, and student and community volunteers. Members of USGBC include engineers, architects, building owners, contractors, facility managers, building product manufacturers, designers, utility managers, college students, college faculties, government agencies, and more.

USGBC promotes buildings that are environmentally responsible, profitable, and supportive of human health.

USGBC's Mission:

"To transform the way buildings and communities are designed, built, and operated, enabling an environmentally and socially responsible, healthy, and prosperous environment that improves the quality of life."

USGBC's Vision:

"Buildings and communities will regenerate and sustain the health and vitality of all life within a generation."

USGBC accomplishes this mission through the following:

- LEED: The most widely known and used green building program across the world.
- Credentials: The LEED Green Associate credential and the various LEED AP credentials offer professionals a designation to help them stand out in the building industry.
- Greenbuild International Conference & Expo: The world's largest expo, launched in 2002, dedicated to green buildings.
- Education: USGBC provides educational programs about green buildings to further spread environmental principles and design.
- Advocacy: At every level of government, USGBC provides policymakers and community leaders with all the tools, strategies, and resources necessary to inspire action toward a sustainable built environment.
- Chapters & Committees: USGBC maintains a network of regional chapters and community organizations across the country for involvement from the local level to the national level.

USGBC has established **six strategic goals** to guide its efforts toward transforming the built environment:

- Sustainable Cities and Communities: Lead the building sector's participation in the broader movement to create sustainable, resilient cities and communities.
- Climate and Natural Resources: Drive the significant reduction and eventual elimination of the building industry's contribution to climate change and natural resource depletion, both during construction and operations.
- Green Building Marketplace: Accelerate the demand for green buildings, make green building practices more accessible, and expand the delivery of sustainable projects across the market.
- Public Policy: Advocate for comprehensive green building policies and codes at the local, state, national, and international levels of government.
- International: Advance green building globally by building certification capacity, sharing knowledge, and collaborating on regionally appropriate practices and policies.
- Organizational Excellence: Strengthen USGBC's internal structure and capacity to support the market transformation needed to achieve its mission.

ABOUT LEED

Leadership in Energy and Environmental Design, in short LEED, is a framework for identifying, implementing, and measuring green buildings, neighborhood, cities and communities design, construction, operations, and maintenance.

LEED is a voluntary, consensus-based framework which serves as a guideline and assessment mechanism for the design, construction, and operation of high-performance, green buildings, neighborhoods, cities and communities.

LEED aims to make the best use of natural resources, encourage strategies that restore and regenerate the environment, reduce the negative impacts of buildings on both human health and the planet, and aims for high-quality indoor environments where occupants can thrive. The system promotes an integrated approach to design, encourages the adoption of proven technologies alongside cutting-edge strategies, and works to advance the expertise of building professionals in sustainable practices.

LEED is a market-driven rating system built on consensus among industry stakeholders. It strikes a balance between requiring established best practices and encouraging the adoption of leadership-level strategies that push the industry forward. The benchmarks LEED sets are intentionally challenging but achievable, defining what it means for a building to be truly green.

The creation of the LEED green building rating system is led by USGBC member-based volunteer committees, subcommittees, and working groups in conjunction with USGBC staff. After development, LEED's green building rating systems are subject to review and approval by the LEED steering committee and the USGBC board of directors prior to a vote by USGBC membership.

With LEED, USGBC strives to:

- Promote the environmental, economic, health, and social benefits of green buildings throughout their entire life cycle.
- Achieve successful launches of new LEED rating systems and products that meet the evolving needs of the market.
- Earn widespread endorsement from private and public real estate industry leaders and stakeholders.
- Gain adoption by federal, state, and local governments as a vehicle for green building policy development and implementation.
- Deliver professional and timely customer service tailored to the needs of project teams and building professionals.
- Offer a comprehensive portfolio of programs that serves the diverse needs of the real estate and construction industry.
- Develop innovative technical tools and support services to help project teams navigate the LEED certification process.
- Establish LEED as the standard for green building practices across homes, commercial buildings, and developments throughout the United States.
- Support international adaptations of LEED in collaboration with green building councils and organizations around the world.
- Lead the industry in practical knowledge about the latest innovations in sustainable design, construction, and operations.

- Continuously improve LEED performance criteria as the industry gains more experience with integrated design, green construction, and sustainable building operations.

LEED CERTIFICATION BENEFITS

LEED-certified buildings are proven to be environmentally friendly and to respect human health. However, there are other major benefits of a LEED certification that can be summarized as money savings over the lifetime of a project, increased project value, and increased building occupant satisfaction.

Additionally, LEED projects are known to be much more energy efficient compared with regular buildings and to have lower operating costs. The same thing applies for water consumption, as LEED-certified buildings consume less potable water than regular buildings. LEED-certified projects have an increased building value, which means more return on investment. LEED is a great marketing tool for developers, which results in high occupancy rates. Since the building systems in LEED-certified projects are commissioned during the design and construction phases as a requirement, the building systems create fewer problems during the operations phase. Furthermore, people who work in LEED-certified buildings have increased productivity due to the healthier environment of the buildings, which is the result of increased exposure to natural daylight, increased ventilation rates inside the building, less contaminated indoor air, more open spaces, increased acoustic performance, and many other advantages. In addition, LEED-certified buildings may qualify for tax rebates, zoning allowances, and other incentives in many cities.

Even though the initial cost of a LEED-certified building may be slightly higher than a regular building, the life-cycle cost of a regular building typically turns out to be much higher than a LEED-certified building.

GBCI

Green Business Certification Inc. (**GBCI**), formerly known as Green Building Certification Institute (GBCI), changed its name on April 16, 2015; is the sister organization of USGBC that administers LEED building certifications and LEED professional accreditations.

When a project submits all the necessary documents for a LEED application, GBCI is the organization that will review the application and determine if all the requirements set forth in a LEED rating system are fully met or not. And if the requirements are fulfilled, GBCI will grant the LEED certificate.

LEED professional exams are also administered by GBCI, and LEED professional credentials are awarded according to the exam results.

LEED PROFESSIONAL CREDENTIALS

LEED credentials offer professionals a designation to help them stand out in the building industry. Currently, there are three main LEED professional credentials:

- **LEED Green Associate**: GBCI has created the LEED Green Associate credential for professionals who want to demonstrate green building knowledge and expertise in nontechnical fields of practice, to denote basic knowledge of green design, construction, and operations. LEED Green Associates have a proven, up-to-date understanding of green building principles and practices. **The LEED GA abbreviation is disapproved by USGBC and cannot be used to demonstrate the LEED Green Associate professional credential**.
- **LEED Accredited Professional with specialty**: The LEED AP with specialty credential is created for professionals with advanced knowledge in green building practices and with specialization in a particular LEED rating system. The LEED AP credential is divided into different specialties and each credential is designed to show in-depth knowledge of a particular rating system. Below is the list of all LEED AP specialties:

  a. LEED AP Building Design + Construction (LEED AP BD+C)
  b. LEED AP Interior Design + Construction (LEED AP ID+C)
  c. LEED AP Operations + Maintenance (LEED AP O+M)
  d. LEED AP Neighborhood Development (LEED AP ND)
  e. LEED AP Homes

  There is also a LEED AP credential without any specialty, called LEED AP or Legacy LEED AP, which was granted to professionals who passed the LEED AP exam before 2009. The Legacy LEED AP credential is not offered anymore, and all the new LEED APs should choose a specialty.
- **LEED Fellow:** According to USGBC, LEED Fellow designates the most exceptional professionals in the green building industry, and it is the most prestigious designation awarded. The program was established to recognize outstanding LEED APs who have demonstrated exceptional LEED-related impacts for at least 10 years.

Other than those mentioned credentials, there are also LEED professional certificates such as LEED Green Rater and Green Classroom Professional. **LEED Green Raters** provide in-field verification to every LEED Residential project. (The other rating systems do not require any in-field verification.) The **Green Classroom Professional** (**GCP**) certificate is designed for educators and school facility professionals who want to gain a foundational understanding of green building principles in the context of a school setting, helping them identify what supports or impedes healthy, resource-efficient, and environmentally sustainable learning spaces.

For exam purposes, it's essential to know that the credentialing exams and building certifications are administered by GBCI, not USGBC. LEED Reference Guides, LEED rating systems, and LEED education programs are developed and managed by USGBC. LEED credentials are for people; LEED rating systems are for buildings/neighborhoods. Professionals are accredited; buildings are certified.

CREDENTIAL MAINTENANCE PROGRAM (CMP)

LEED Green Associates and LEED APs with specialty should maintain their credentials by meeting the requirements of the Credential Maintenance Program (CMP) every two years, or otherwise the credential will expire, and retaking the exam will be necessary to maintain the LEED professional credential.

Under the CMP, all LEED Green Associates must earn **15 continuing education (CE) hours every two years,** and LEED APs must earn **30 continuing education (CE) hours every two years** after earning their credential.

END OF CHAPTER EXERCISES

EXERCISE: MATCHING GAME

Match the following terms to their definition.

Terms		
USGBC's Mission	GBCI	LEED Accredited Professional with Specialty
USGBC's Vision	LEED Green Raters	
LEED Green Associate	Green Classroom Professional (GCP) Certificate	

1) ____________________________: Designed for educators and school facility professionals who want to gain a foundational understanding of green building principles in the context of a school setting, helping them identify what supports or impedes healthy, resource-efficient, and environmentally sustainable learning spaces.

2) ____________________________: LEED credential created for professionals with a proven, up-to-date understanding of green building principles and practices.

3) ____________________________: LEED credential, created for professionals with advanced knowledge in green building practices and specialized in a particular LEED rating system.

4) ____________________________: The sister organization of USGBC that administers LEED building certifications and LEED professional accreditations.

5) ____________________________: Professional who provides in-field verification to LEED Residential projects.

6) ____________________________: To transform the way buildings and communities are designed, built, and operated, enabling an environmentally and socially responsible, healthy, and prosperous environment that improves the quality of life.

7) ____________________________: Buildings and communities will regenerate and sustain the health and vitality of all life within a generation.

ANSWER: MATCHING GAME

1) Green Classroom Professional (GCP) certificate	4) GBCI	7) USGBC's Vision
2) LEED Green Associate	5) LEED Green Raters	
3) LEED Accredited Professional with Specialty	6) USGBC's Mission	

PRACTICE EXAM

To access the end-of-chapter practice exam online in our exam simulator, scan the QR code below with your laptop, phone, tablet, or another compatible device, then log in using your email address and password. If you have trouble scanning the QR code, visit www.projectific.com/account and log in using your email address and password. If you have not yet created a password, please refer to the *How to Access the Exam Simulator* section at the back of the book for instructions.

CHAPTER 3: LEED GREEN BUILDING CERTIFICATION PROGRAM

This chapter will explain how a project can receive LEED certification, what the requirements and processes are, and which types of LEED certifications are available. Let's take a closer look at the structure of LEED.

LEED STRUCTURE OVERVIEW

Just like a car rental company categorizing its rental vehicles in categories such as SUVs, minivans, compact cars, standard-sized cars, etc., LEED categorizes projects under **LEED rating systems** according to their types of construction and space usage. For example, a newly built office project may fall under the LEED Building Design + Construction rating system, but a hotel interior renovation project would fall under the LEED Interior Design + Construction rating system.

The first thing a project team should do is to choose the LEED rating system that's appropriate for its project before registering the project for the LEED certification process. Then the project team can proceed with registering its project at Arc, the online platform where project teams manage their LEED certification from registration through documentation and review. (Arc will be discussed in detail in the following sections.)

No one can blame a sports car for not handling the road properly on a dusty mountain road. However, an SUV would be required to perform well on mountain roads. Just like we have different expectations from each vehicle category, LEED has different expectations from each rating system in terms of awarding the LEED certification. To summarize, after a rating system is chosen, project teams should basically need to work on fulfilling its requirements.

Let's take a look at the first group of requirements, which are applicable to all the LEED rating systems. In other words, they're applicable to all the projects pursuing LEED certification. They're called the Minimum Program Requirements (MPR). These will also be discussed in the following sections in more depth, but to summarize, a project must be in a permanent location on existing land, must use reasonable LEED boundaries, and must comply with LEED's project-size requirements. Without fulfilling those three MPRs, a project cannot qualify for a LEED certification, and this is applicable for all ratings systems—period.

After the project complies with the MPRs, it's time to comply with the requirements of the selected rating system. There are two things a project should accomplish under each rating system. The first thing is to satisfy all the prerequisites of that rating system. You can think about them as if they're MPRs inside each rating system. Without achieving all of the prerequisites, a project cannot achieve a LEED certification. The second thing a project should do is collect points by achieving the credits of that rating system. A minimum of 40 points out of 100 is required for a LEED certification. (All LEED rating systems are scored on a basis of 100 total points, and 40 points are the minimum number of points to become LEED Certified, which is the base level.) Please note that satisfying prerequisites does not result in the award of any points; only credits do.

Every credit has a different requirement, and every credit awards a point or points. Project teams decide on choosing the credits they want to pursue, preferably during the beginning of the project. There are approximately 50 credits under the LEED Building Design + Construction (LEED BD+C) rating systems and each has different point values. After achieving the chosen credits, the projects need to prepare documentation to show that they achieved their MPRs, the prerequisites, and enough credits, and basically upload that documentation to Arc. Done!

To summarize, in order to achieve a LEED certification, first there are the MPRs (just three), which every project must comply with regardless of the rating system. Then there are prerequisites of each rating system that are mandatory to achieve for LEED certification. Then the project team must choose credits to pursue and collect more than 40 points. And all should be documented. Don't forget, all the MPRs and prerequisites are mandatory. However, projects are free to choose the credits they want to pursue.

What if a project satisfies all the MPRs and prerequisites and gets a total of 70 points? Then it will be more than LEED Certified and will be given a LEED Gold certificate!

According to the total number of points a project receives, LEED certification levels are determined:

- **LEED Certified™: 40–49 points** (This is the base certification level.)
- **LEED Silver: 50–59 points**
- **LEED Gold: 60–79 points**
- **LEED Platinum: 80–80+ points**

(Knowing the points required for each level of LEED certification is essential for the exam.)

However, if a project achieves all the MPRs and earns 90 points but misses only one prerequisite, no LEED certificate will be awarded.

To see the complete list of prerequisites/credits and the point values for credits, it's recommended that you review the LEED scorecard for each rating system from USGBC's website. The **LEED scorecard** allows project teams to select the credits they plan to pursue and determine whether the total expected points will be sufficient to achieve the targeted level of LEED certification.

For all the LEED BD+C rating systems, the prerequisites and credits are categorized under eight categories, which are called the **credit categories.** (Each credit category will be discussed in depth, starting with Chapter 4.) Following are the credit categories:

- Integrative Process, Planning and Assessments (1 point)
- Location and Transportation (15 points)
- Sustainable Sites (11 points)
- Water Efficiency (9 points)
- Energy and Atmosphere (33 points) **(Category containing the highest points!)**
- Materials and Resources (18 points)
- Indoor Environmental Quality (13 points)
- Project Priorities (10 points)

Did you notice that there are 110 total points available? Just like some exams have bonus questions, LEED has 10 bonus points. Project Priorities credit category award the bonus points in LEED. But still, every project will be scored on the basis of 100 points, not 110.

LEED V5 RATING SYSTEMS

Just like individuals have credentials such as LEED Green Associate or LEED AP BD+C, all LEED projects are certified under a single LEED rating system.

All the LEED v5 rating systems are grouped under three broad categories and are then further organized by project type and scope as follows:

1) LEED for Building Design and Construction (LEED BD+C) rating systems

a) LEED BD+C: New Construction and Major Renovations
b) LEED BD+C: Core and Shell Development

2) LEED for Interior Design and Construction (LEED ID+C) rating systems

a) LEED ID+C: Commercial Interiors

3) LEED Building Operations and Maintenance (LEED O+M) rating systems

a) LEED O+M: Existing Buildings

Note that the entire gross floor area of a LEED project must be certified under a single rating system—such as LEED BD+C: New Construction and Major Renovations or LEED ID+C: Commercial Interiors—regardless of mixed construction or space usage type. The following sections describe each rating system in detail.

LEED FOR BUILDING DESIGN AND CONSTRUCTION (LEED BD+C) V5 RATING SYSTEMS

All types of buildings—whether newly constructed or existing buildings undergoing major renovations—will fall under the LEED BD+C category. Major renovations are defined as buildings with extensive interior alteration work in addition to work on the exterior shell and/or primary structural components and/or the core and peripheral mechanical, electrical, plumbing, and service systems and/or site work.

1) LEED BD+C: New Construction and Major Renovations

This is the most widely used rating system under LEED BD+C. It addresses design and construction activities for both new buildings and major renovations of existing buildings. To be eligible, **at least 60% of the project's gross floor area must be complete** by the time of LEED certification. **Complete spaces** are those that have their basic floor, wall, and ceiling finishes installed along with essential mechanical, electrical, and plumbing systems or fixtures necessary to occupy the space for its intended use.

2) LEED BD+C: Core and Shell Development

This is for projects in which the developer controls the design and construction of the building core and the entire mechanical, electrical, plumbing, and fire protection systems, but does not complete the interior tenant fit-out. An example would be an office building in which the common spaces are constructed by the developer and the tenant spaces are left incomplete, allowing the tenants to construct their own custom designs in their office spaces.

To be eligible for this rating system, **at least 40% of the project's gross floor area must be incomplete** at the time of certification. **Incomplete spaces** are those that do not have their basic floor, wall, and/or ceiling finishes installed or essential mechanical, electrical, or plumbing systems or fixtures necessary to occupy the space for its intended use.

LEED FOR INTERIOR DESIGN AND CONSTRUCTION (LEED ID+C) V5 RATING SYSTEM

LEED ID+C rating system applies to interior renovation projects of commercial spaces that have **no scope for the exterior shell of the building or site work**. The team installs movable furnishings, fixtures, and equipment (FF&E) to support the intended regular operations of the space.

1) LEED ID+C: Commercial Interiors

This is for interior spaces undergoing a complete fit-out. If a company rents an office in a building and decides to design and construct its interior office space, the LEED ID+C: Commercial Interiors rating system can be used to certify that space. Think about a LEED-certified Core and Shell office building in which the developer was not involved in the design and construction of the tenant spaces. In this scenario, the tenants may decide to pursue a LEED ID+C: Commercial Interiors certification for their office spaces, which would only cover the interior fit-out design and construction work.

LEED BUILDING OPERATIONS AND MAINTENANCE (LEED O+M) V5 RATING SYSTEM

The LEED O+M rating system applies to existing buildings that are focused on operational improvement. This rating system covers whole buildings that have been **fully operational and occupied for at least one year**. Rather than evaluating design and construction, LEED O+M certifies the ongoing operations and maintenance of the building through performance-driven strategies and outcomes, allowing buildings in use to achieve greater efficiency. LEED O+M monitors the building's systems—HVAC, electrical, automation systems, and more—as well as the building's overall performance.

A LEED ID+C: Commercial Interiors certified office located in a LEED BD+C: Core and Shell certified building can also receive a LEED O+M: Existing Buildings certification. By earning a LEED O+M: Existing Buildings certification, the project demonstrates that its ongoing building operations are efficient and high in performance.

It is important to remember that LEED O+M certifications expire every three years, and in order to maintain a LEED O+M certification, a project needs to be **recertified every three years**. Certifications under the other rating systems do not expire and do not require recertification.

1) LEED O+M: Existing Buildings

This is the primary rating system under LEED O+M and applies to existing buildings of all use types that are focused on improving operational performance.

LEED FOR CITIES V5 RATING SYSTEM

At the time of writing this book, LEED v5 Cities is under development. LEED for Cities is a rating system designed to help cities and communities measure, manage, and improve their sustainability performance at a broader scale. Rather than certifying a single structure, LEED for Cities evaluates the environmental, social, and economic performance of an entire city or community.

PROJECT TYPE REQUIREMENTS

Once the rating system is selected, the project team will work on fulfilling the prerequisites and credits that are specific to that rating system. In other words, all projects registered under the LEED BD+C: New Construction and Major Renovations rating system will be subject to the same prerequisites and credits.

However, some prerequisites and credits also define specific requirements for specific project types, such as healthcare, schools, warehouses and distribution centers, or residential projects. For example, the Compact and Connected Development credit under the LEED BD+C: New Construction and Major Renovations rating system has specific requirements for schools, data centers, warehouses, distribution centers, and healthcare projects. In such instances, mixed-use projects should use the 40/60 rule to identify the appropriate project type.

Under the **40/60 rule**, a project should be divided into sections according to the appropriate project type for each section. Then, the total floor area corresponding to each project type should be calculated. If the total floor area of one of the applicable project types is less than 40% of the project's total floor area, that project type cannot be used. If the total floor area of one of the applicable project types is more than 60% of the project's total floor area, that project type must be used. If it falls between 40% and 60%, then the project team can decide which project type to use.

Let's consider a twenty-story high-rise building project with equal floor area on each floor, which contains retail units on the first five floors and residential units on the floors above. Since 75% of the project's total floor area is residential, the project team will follow the residential project type requirements in the prerequisites and credits where they exist.

MINIMUM PROGRAM REQUIREMENTS (MPR)

The **minimum program requirements** are the minimum standards/characteristics that every LEED project should meet to pursue LEED certification. MPRs state whether a project can pursue LEED certification or not. They provide guidance on the types of projects that are eligible for LEED certification, protect the integrity of the LEED program, and reduce the number of issues that arise during the certification process. They define the types of buildings, spaces, and neighborhoods that a LEED rating system is designed to evaluate. (It is very important to know the MPRs for the exam purposes.)

There are three MPRs:

1) Every project must be in a permanent location on existing land

A project that is designed to move at any point in its lifetime is not eligible for LEED certification. Since a significant portion of LEED's prerequisites and credits are dependent on the project's location, the certification is awarded according to that particular location. Prefabricated or modular structures can be LEED-certified as long as they are installed permanently.

It is also important to locate the projects on existing land to avoid artificial landmasses, which can displace and disrupt ecosystems in the future.

2) Must use reasonable LEED boundaries

The LEED project boundary must include all contiguous land that is associated with the project and that supports its typical operations. These will include the landscaping, septic or stormwater treatment equipment, parking, sidewalks, and even more. Any project space cannot be shown to be excluded in order to give the project an advantage in complying with credit requirements.

The LEED boundary may not unreasonably exclude portions of the building or site to give the project an advantage in complying with prerequisite and credit requirements. The **LEED project boundary** is defined by the platted property line of the project, including all land and water within it.

3) Must comply with project size requirements

All LEED projects must meet the following size requirements according to their rating system:

- LEED BD+C and LEED O+M rating systems: The project must include a **minimum of 1,000 square feet (93 square meters) of gross floor area.**
- LEED ID+C rating systems: The project must include a **minimum of 250 square feet (22 square meters) of gross floor area.**

PREREQUISITES AND CREDITS

Whether deciding to pursue LEED certification or not, all project teams should first take a look at the MPRs to see if their project can be eligible for a LEED certification. Then the project teams will need to determine the rating system that their project fits into. After the rating system is selected, project teams should evaluate the prerequisites and credits of that rating system.

Prerequisites are the minimum requirements that all buildings under a certain rating system must meet in order to achieve LEED certification. For example, all the projects that are certified under the LEED BD+C: New Construction and Major Renovations rating system must meet all the prerequisites of that rating system, while a project registered under LEED O+M: Existing Buildings must meet all of its prerequisites. To illustrate, let's take a look at the one of the prerequisites under the LEED BD+C: New Construction and Major Renovations rating system.

Prerequisite—Fundamental Refrigerant Management: This prerequisite requires projects to either use no refrigerants in their refrigerant-containing equipment or to use refrigerants with low global warming potential (GWP).

Therefore, every project under the LEED BD+C: New Construction and Major Renovations rating system must either avoid refrigerants entirely or select refrigerants with low GWP—period. Any project that cannot meet this requirement will not be eligible for a LEED certification. While projects can pick and choose the credits they want to pursue, prerequisites are mandatory. Fulfilling the requirements of prerequisites will also not result in the award of any points.

Now let's look at the credits, which are also a part of each rating system. Project teams are free to go for any credit they want within the selected rating system. Each **credit** has requirements to be fulfilled, and once the requirements are met, the project will earn points. The higher the total points a project earns, the higher the level of LEED certification that will be awarded. To illustrate how a credit works, let's take a look at the one of the credits under the LEED BD+C: New Construction and Major Renovations rating system.

Credit—Compact and Connected Development: This credit rewards projects that are located in compact, walkable, mixed-use areas with access to diverse uses and quality transit, reducing greenhouse gas emissions resulting from single-occupancy vehicle use and encouraging more sustainable transportation patterns.

Hence, any LEED BD+C: New Construction and Major Renovations project located in a compact, well-connected area with access to quality transit and diverse uses can earn points under this credit. Projects that do not comply with this credit will not receive any points. As mentioned, there are **100 total points** allocated to the credits under each rating system; plus, there are **10 bonus points.** A project that earns 50 points total will receive a LEED Silver while another project that earns 70 points will receive a LEED Gold (if both of them satisfied all the prerequisites and the MPRs).

The prerequisites and credits are also sometimes categorized as **design prerequisites/credits** and **construction prerequisites/credits,** according to the project phase they are performed in. For example, the previously mentioned Compact and Connected Development credit is a design credit since the project's location will not be decided during the construction phase.

PILOT CREDITS

Other than the prerequisites and credits, there are also **pilot credits**, which are tested across all rating system types and credit categories and include credits proposed for the next version of LEED. Project teams can pursue any of these pilot credits under Project Priorities credit and earn points by providing USGBC with feedback on the credit. Based on the feedback USGBC receives from projects that pursue these credits, pilot credits may become official credits in the next LEED rating system.

The pilot credits available to each rating system can be found from the **LEED Pilot Credit Library at;**

- http://www.usgbc.org/pilotcredits

LEED IMPACT AREAS AND CREDIT POINT ALLOCATION PROCESS

By looking at the LEED BD+C: New Construction and Major Renovations scorecard, it can be seen that each credit contains a different number of points. For example, the Enhanced Energy Efficiency credit contains 10 points while the Sensitive Land Protection credit contains only 1 point.

To better understand the system that LEED uses to distribute points to each credit, we first need to understand the impact areas that shape the entire LEED v5 framework.

LEED v5 is structured around **three impact areas** which are the overarching objectives that every LEED project aims to accomplish, and every prerequisite and credit in the rating system is connected to one or more of them. (Knowing the impact areas is essential for the exam.)

The three impact areas are:

1) Decarbonization (50% of available points): This is the dominant impact area in LEED v5, reflecting the urgency of the climate crisis and the building industry's significant contribution to greenhouse gas emissions. Decarbonization targets all major sources of carbon emissions from buildings: operational carbon from energy use, embodied carbon from materials and construction, refrigerant emissions from mechanical systems, and transportation emissions from occupant commuting. LEED v5 enhances the carbon literacy of project teams through new prerequisites that require operational carbon projections and decarbonization plans, providing a visual forecast of how a building's carbon emissions are expected to change over time as the electrical grid becomes cleaner. Key strategies include improving building envelopes, electrifying heating and cooling systems, reducing peak thermal loads, enhancing energy efficiency, and procuring renewable energy. On the materials side, project teams are encouraged to minimize embodied carbon through whole-building life-cycle assessments, analysis of environmental product declarations (EPDs), jobsite emissions tracking, waste diversion, and circular economy practices that reduce demand for virgin resources.

2) Quality of Life (25% of available points): This impact area recognizes that buildings are fundamentally about people. It addresses the health, well-being, resilience, and equity of building occupants and surrounding communities. LEED v5 requires project teams to understand who uses the building, assess the needs of the local community, evaluate natural hazards that could affect the site, and consider the workers and supply chains behind building materials. The framework rewards strategies that promote adaptable and resilient design, protecting both occupants and communities from climate disruptions. It also encourages enhancing the economic and social vitality of communities by strengthening local workforces and expanding access to both indoor and outdoor spaces. Credits address a wide range of human experiences—from indoor air quality, thermal comfort, daylight, and acoustics to inclusive design, occupant well-being, and biophilic connections with nature—fostering spaces that are healthy, comfortable, equitable, and enjoyable.

3) Ecological Conservation and Restoration (25% of available points): This impact area focuses on limiting the environmental degradation caused by building construction and operations and promoting the rehabilitation and restoration of ecosystems. LEED v5 takes a comprehensive approach to ecological health by incorporating strategies that conserve natural habitats, minimize impacts on land and wildlife, restore ecosystems to enhance resilience and biodiversity, and manage natural resources responsibly. Credits reward projects for protecting sensitive lands, increasing density in already developed areas to reduce sprawl, implementing low-impact development practices to prevent flooding and erosion, restoring soils, and using native and adaptive plants. Water conservation plays a central role, with strategies for reducing consumption, submetering usage, and detecting leaks to prevent waste and protect water resources. On the materials side, prioritizing reuse, diverting waste from landfills, and supporting circular economy practices decrease reliance on virgin materials, reduce methane emissions, and preserve the natural resources that ecosystems depend on.

Each credit in LEED makes a contribution to one or more of these impact areas. Credits that most directly address the prioritized impacts are given the greatest weight—meaning they contain more points. For example, a credit that significantly reduces operational carbon emissions will carry more points than a credit with a narrower environmental benefit. This weighting process ensures that the point structure drives project teams toward the strategies that matter most for climate action, human well-being, and ecological health.

HOW LEED IS DEVELOPED

LEED is developed through a transparent, consensus-based process by volunteer committees composed of USGBC members representing a balanced cross-section of stakeholders across the building industry. This development process is guided by three key principles: transparency, openness, and inclusiveness.

The improvements to LEED are developed by these volunteer committees, subcommittees, and working groups, in conjunction with USGBC staff. The proposals are then reviewed and affirmed by the LEED Steering Committee (LSC) and the USGBC Board of Directors before going to a vote by the relevant consensus committee and ratification by USGBC's membership.

LEED COMMITTEES

USGBC relies on a structure of volunteer committees to develop and maintain LEED. The key groups involved are:

- The **LEED Steering Committee (LSC)** is a standing committee established by the USGBC Board of Directors. The LSC sets the strategic direction for LEED development, ensures consistency among all LEED rating systems, and reviews recommendations from other committees to confirm alignment with the overall LEED framework. Four principles guide the LSC: transparent governance, technical rigor paired with market relevance, consistency across all rating systems, and collaboration among committees and staff. The LSC is made up of volunteer voting members and USGBC staff as non-voting members.

- **Consensus Committees** are established by the USGBC Board of Directors to support the creation and updating of specific LEED rating systems. Each consensus committee is structured to represent different technical and market interests, and no single interest group may make up more than 25% of the voting membership. The seven interest groups represented include building owners and developers; architects, engineers, and designers; contractors and facility operators; product manufacturers; building occupants and community advocates; regulation and standards organizations; and utilities, academia, finance, and others. Consensus committees vote on the final language of LEED rating systems before they are published for member ratification.
- **Technical Advisory Groups (TAGs)** advise the LSC and consensus committees on specific technical aspects of LEED, such as materials or indoor air quality.
- **Working Groups** come together for the completion of a specific task with a clearly defined scope and timeline. They are established by the LSC or a consensus committee.
- **Advisory Task Forces** are short-duration groups designated by staff to informally advise on issues limited in scope during the development process.

LEED RATING SYSTEM EVOLUTION

LEED is updated through a regular development cycle. USGBC has committed to updating LEED **every five years**, and there are three basic ways the rating system is improved:

1) Implementation and maintenance of current version: LEED rating systems are continually improved through published updates such as addenda. Addenda is the overarching term for all changes made to a LEED rating system between major version releases. Addenda include corrections, clarifications, LEED interpretations, and alternative compliance paths (ACPs). LEED interpretations are precedent-setting official answers to formal inquiries about how to implement LEED requirements on a project. **Alternative compliance paths (ACPs)** provide additional options for meeting credit requirements, particularly for international projects that need to reference local or regional standards instead of United States–based standards. (Addenda and LEED interpretations will be discussed in detail in the following sections.)

2) LEED rating system adaptations: Specific project type requirements are built into the prerequisites and credits of the main rating systems, addressing the unique needs of project types such as schools, healthcare, data centers, warehouses, and residential projects. Alternative compliance paths (ACPs) also serve an important role by enabling international projects to meet local or regional standards instead of United States–based standards where appropriate. There are two types of ACPs: **Regional ACPs**, which are credit requirements that may optionally be applied in specific regions of the world, and **pilot ACPs**, which are paths USGBC is considering for inclusion in future versions of LEED and can optionally be used in place of existing LEED credits.

3) Next version of LEED: The most comprehensive type of development is the creation of a new version of the rating system. This process includes multiple avenues for stakeholder input, public comment periods, consensus committee approval, and ratification by USGBC's membership. Anyone can propose changes to LEED at any time, and USGBC maintains a record of all proposals for review by the LEED Steering Committee and the relevant consensus committees.

Once a draft is ready, it is published for public comment, revised based on feedback, and then submitted for consensus committee approval and member ratification.

Pilot credits also plays an important role in the evolution of LEED. A pilot credit must be at least one year old before its results are evaluated, and the relevant consensus committee then decides whether to close the pilot, keep it in the pilot credit library, or incorporate it into the rating system.

USGBC also uses beta versions to test changes to an entire rating system before it goes through public comment and balloting. Real projects participate in the beta versions, follow the same submittal and review process as any other active version, and can earn certification if achieved.

LEED INTERPRETATIONS

When a project team needs clarification about the technical details of a prerequisite, credit, or MPR, a **LEED interpretation** may be issued to ask for clarification. Unlike other forms of guidance, LEED interpretations are **precedent-setting**. This means that once a LEED interpretation is published, it applies to all LEED projects—not just the one that requested it. The clarification effectively becomes part of the rating system itself.

LEED interpretations are published online in the Addenda database on USGBC's website for general access and are subject to a consensus-based review by USGBC. Projects are required to adhere to LEED interpretations posted before their registration date. Adherence to interpretations posted after a project registers is optional but encouraged.

LEED interpretations are administered by USGBC, not GBCI. (That distinction is important to know for exam purposes.)

LEED interpretations can become a part of addenda, which will be the next topic.

ADDENDA

Addenda is the overarching term that encompasses all changes, improvements, and modifications made to a LEED rating system. Addenda includes four types of changes: corrections, clarifications, LEED interpretations, and alternative compliance paths (ACPs). Think of addenda as the mechanism through which LEED keeps itself current between major version releases—similar to how a piece of software receives periodic updates to fix issues and add new features.

Addenda include both substantive and non-substantive changes. **Substantive changes**—such as corrections that fix inaccuracies, LEED interpretations that expand upon existing language, and ACPs that provide additional compliance options—must be brought to the appropriate LEED committee for review and recommendation before taking effect. **Non-substantive changes**, such as grammar, spelling, and formatting corrections, do not require committee input.

A project that registers after any issued addenda will be automatically subject to the requirements of those addenda. Projects are held to the published addenda in place at the time of their registration. However, project teams are free to apply any addenda released after their registration date and are encouraged to do so to stay current with the latest guidance.

Addenda for LEED v5 are expected to be released **twice per year**. However, LEED resources, pilot credits, innovation credits, ACPs, Regional ACPs, and project type–specific credits may be added or updated at any time through the Project Priority directory and the LEED credit library. Addenda can be found in the Addenda database on USGBC's website, which is searchable by rating system, credit category, and project location.

LEED REFERENCE GUIDES

LEED reference guides are the comprehensive manuals of LEED, created by USGBC, which defines all the instructions, prerequisite/credit intentions, prerequisite/credit requirements, term definitions, required documentation, and other information necessary to LEED-certify a building.

There is a LEED reference guide for each broad category of the rating systems:

- LEED v5 Reference Guide for Building Design and Construction
- LEED v5 Reference Guide for Interior Design and Construction
- LEED v5 Reference Guide for Building Operations and Maintenance

LEED CERTIFICATION PROCESS FOR LEED BD+C AND LEED ID+C PROJECTS

1. PROJECT REGISTRATION

The first step for LEED-certifying a building is to register the project through Arc, which means completing the registration form and paying the **registration fee** that varies by membership status. In this phase the project team should select the appropriate LEED rating system.

The certification platform defines two key roles: the owner and the LEED Project Administrator. The **owner** of the project refers to the individual(s) or entity(ies) that has the legal authority to fulfill the obligations defined in the LEED Certification Agreement. The owner has ultimate control over the LEED certification application, meaning that GBCI will respond to the owner regarding the administration of the project over any other member of the project team.

As discussed, a **LEED Project Administrator** is the primary project contact for GBCI. The LEED Project Administrator is the team member who acts as a project manager, overseeing the LEED project and organizing team members for certain tasks, credits, or prerequisites. The project administrator makes sure the LEED submission is complete and accurate before submitting the project to GBCI for review, and accepts the review results once the review is complete.

Note that the individual who initially registers the project will automatically be granted the role of the LEED Project Administrator, but this role can be transferred to another team member at any time by the administrator or owner.

2. APPLICATION

After the project registration phase, the project team will identify LEED credits to pursue and assign responsibility for the various credits to project team members. The team will then collect information, perform calculations, prepare documentation, fill out the credit forms demonstrating the achievement of the prerequisites and credits, and upload all the documentation to Arc. Once everything is complete, the LEED Project Administrator will verify the submission and the project team will pay the review (certification) fees to submit the application.

Review (certification) fees are based on the gross floor area of the project and membership status; unlike the registration fee, they are not a flat fee. Once the review fees are received, GBCI will begin the project review process. Expedited review is also available for an extra flat fee.

If a project team has questions about the technical details of a prerequisite, credit, or MPR, a **credit interpretation ruling (CIR)** is issued to ask for clarification. Sometimes the reference guide will not be able to provide all the information regarding the achievement of a particular prerequisite/credit, and in that case, the project team would submit a CIR and ask for a clarification. CIRs are administered by the GBCI, and the Credit Ruling Committee reviews the CIR.

Note that LEED v4 and v4.1 BD+C projects and LEED v4.1 O+M projects are also eligible for precertification. (At the time of writing this book, precertification is not yet available for LEED v5.) **Precertification** is an optional review pathway that helps project teams determine which credits and prerequisites the project is likely to achieve during the full review. It also allows developers to market the unique and valuable green features of the proposed building to potential tenants and financiers. Precertification is valid for **three years**. However, it does not guarantee LEED certification, since the final certification will be awarded at the end of construction when all the requirements are met.

3. REVIEW

There are two types of review options for LEED BD+C and LEED ID+C projects. The first option is the **combined review**, in which documentation for all the design and construction prerequisites/credits are submitted for review at the end of the construction phase. At the end of the combined review, GBCI will mark the submitted prerequisites/credits as **awarded** or **denied.** If there is more information needed by GBCI about a prerequisite/credit, GBCI will mark it as **clarify**, and the project teams will submit the requested clarification. It is a little riskier than the second option, which is the **split review.** If some of the expected credits are not awarded, there will not be a makeup since the project is already completed and no change in design or construction can be made. If a prerequisite is not awarded, then the project will not be able to receive a LEED certification at all.

In the second option, **split review**, the design prerequisites/credits are submitted for review during the design phase, and both the additional design prerequisites/credits and all the construction prerequisites/credits are submitted at the end of the construction phase. When the design review is complete, GBCI will either mark the design prerequisites/credits as **anticipated** or **denied.** No prerequisite/credit will be awarded during the design phase since the design will also need to be implemented on-site during the construction phase. If a design prerequisite/credit is marked as anticipated, it means that the project will earn it at the end of the construction phase once that design is implemented on-site. If it's marked as denied, in order to earn the prerequisite/credit, the project teams will need to come up with a design alternative rather than proceeding with that design. At the end of construction phase, the additional design prerequisites/credits (if applicable) and the construction prerequisites/credits are sent for review. GBCI will this time mark all of them as **awarded** or **denied** unless clarification is needed. If more information is needed about a prerequisite/credit, it will be marked as **clarify**, and the project teams will submit the requested clarification.

To summarize the review process, let's list all the steps for a combined review and a split review:

Combined review:

At the end of the construction phase:

1. Both design and construction submittals are sent by the project team.
2. There's a **preliminary review** by GBCI.
3. If additional information is needed to evaluate the application, GBCI will ask for clarifications and the project team will send the clarifications.
4. GBCI will proceed with the **final review** to evaluate the clarifications and to report the awarded/denied prerequisites and credits to the project team.
5. The project team will either accept or appeal the result.

Split review:

During the design phase:

1. Design submittals are sent by the project team.
2. **Preliminary design review** is conducted by GBCI.
3. If additional information is needed to evaluate the application, GBCI will ask for clarification, and the project team will send the clarification.
4. GBCI will proceed with the **final design review** to evaluate the clarifications and report the anticipated/denied prerequisites/credits to the project team.
5. The project team will either accept or appeal the result.

At the end of the construction phase the following occurs:

6. Both construction and extra design submittals (if applicable) are sent by the project team.
7. A **preliminary construction review** is conducted by GBCI.
8. If additional information is needed to evaluate the application, GBCI will ask for clarification, and the project teams will send the clarification.
9. GBCI will proceed with the **final construction review** to evaluate the clarifications and report the awarded/denied prerequisites/credits to the project team.
10. The project team will either accept or appeal the result.

There is no limit on the number of appeals a project can submit, though each appeal requires a fee. (For exam purposes, note that review fees depend on the gross floor area of the project and membership status, while fees for registration, appeals, CIR, and LEED interpretations are flat fees that vary only by membership status, not by gross floor area.)

4. CERTIFICATION OR DENIAL

This is the last step of the LEED certification. GBCI issues the final certification report, which will show the level of LEED certification awarded unless the certification is denied. The project teams can again appeal the results if needed, and if no appeals are made, then the project will be deemed "**closed out,**" which means the project team will no longer be able to appeal the certification level or review decisions for specific credits or prerequisites.

As mentioned, the total number of earned points will determine the level of LEED certification, assuming all the prerequisites and MPRs are met.

- LEED Certified™: 40–49 points
- LEED Silver: 50–59 points
- LEED Gold: 60–79 points
- LEED Platinum: 80–80+ points

This point system applies to all the rating systems in LEED. Any project that earns 70 points will get the LEED Gold regardless of whether it is a LEED O+M project or a LEED BD+C project or belongs to any other rating system. However, LEED Platinum certification has additional requirements beyond earning 80 or more points.

LEED v5 Platinum projects must also meet specific decarbonization requirements in designated credits. For LEED BD+C: New Construction projects, for example, Platinum certification requires meeting minimum thresholds in four specific credits:

- Electrification Credit — Five points are required.
- Enhanced Energy Efficiency Credit — Eight points are required.
- Renewable Energy Credit — 100% of site energy use from any combination of Tier 1, Tier 2, and Tier 3 renewable energy.
- Reduce Embodied Carbon Credit — 20% reduction in embodied carbon.

Core and Shell, LEED ID+C, and LEED O+M projects have their own Platinum thresholds in similar decarbonization-related credits.

LEED CERTIFICATION PROCESS FOR LEED O+M PROJECTS

A LEED O+M certification is not granted for good, unlike the case with the other rating systems. A LEED O+M project needs to get recertification **every three years**. And there isn't a split review option available. The LEED O+M project will only have an initial review, which at the end may result in the grant of the LEED O+M certificate. The project will then have recertification reviews every three years to be able to continue its certification. The rest of the certification process is the same with the LEED BD+C and LEED ID+C projects.

LEED VOLUME AND LEED CAMPUS

LEED Volume Program is for organizations planning to certify a large number of new construction buildings or interior spaces based on a common prototype. It streamlines certification by focusing on similarities in building design and construction practices regardless of location.

LEED Campus Program is designed for multiple buildings on a shared site under the control of a single entity. It allows project teams to document shared site-related credits once and apply them across all individual building certifications on the campus. It simplifies the certification process for campus settings such as universities, hospitals, corporate campuses, and industrial parks.

At the time of writing this book, these programs are not yet available for LEED v5 and are currently being planned by USGBC.

ARC

Arc is the online platform where project teams register their projects for LEED certification, complete prerequisite and credit documentation, and submit applications to GBCI for certification review. It serves as the primary tool for managing the entire LEED certification process from start to finish.

For documenting the LEED prerequisites and credits, project teams fill in the prerequisite and credit forms available on the platform. With the use of standardized credit forms, every LEED project prepares submittals in the same format, which streamlines both the preparation of LEED applications and the review process. For prerequisites and credits that require calculations, calculators are built into the credit forms, which project teams can use to input their data.

All project team members can access the prerequisites and credits uploaded to Arc. GBCI reviews the submitted documents through Arc and grants LEED certification upon successful completion.

USGBC POLICIES AND APPROVED ABBREVIATIONS

When using the USGBC logo, the proportions of the original logo's height and width should remain the same. Scaling and no any other alterations can be made to the logo. USGBC should always be acknowledged with use of the logo. The same conditions apply for the use of the LEED logo, and LEED should be acknowledged.

When using the USGBC name on any first reference, it should be written as "U.S. Green Building Council." In subsequent references, the "USGBC" abbreviation can be used. Other naming or abbreviations cannot be used.

The "LEED GA" abbreviation is used commonly everywhere. However, this abbreviation is disapproved by USGBC and should not be used. Approved usage is "LEED® Green Associate™."

For the "LEED Accredited Professional®" title, the "LEED AP®" abbreviation can be used, followed by the specialty, which can be written as "LEED AP® BD+C" or "LEED AP® Homes."

When describing a LEED certification process, "LEED certification" with lowercase "c" can be used. "LEED-certified" should be used to refer to any project that achieved LEED certification. "LEED Certified" is used to describe a project that has been certified to the base level (for the projects that receive 40–49 points). Projects certified above the base level may be described as "Project X is LEED Gold," "Project X is LEED Silver-certified," or "Project X is LEED-certified at the Platinum level.

END OF CHAPTER EXERCISES

EXERCISE: MATCHING GAME

Match the following terms to their definition.

Terms		
LEED Volume Program	LEED Steering Committee (LSC)	Technical Advisory Groups (TAGs)
LEED Campus Program	LEED Project Administrator	Impact Areas
Alternative Compliance Paths (ACPs)	LEED Interpretation	Pilot Credits

1) ____________________________: A standing committee established by the USGBC Board of Directors. They set the strategic direction for LEED development, ensures consistency among all LEED rating systems, and reviews recommendations from other committees to confirm alignment with the overall LEED framework.

2) ____________________________: The primary project contact for GBCI, overseeing the LEED project and organizing team members for certain tasks, credits, or prerequisites.

3) ____________________________: A precedent-setting official answer to a formal inquiry about the technical details of a LEED prerequisite, credit, or MPR.

4) ____________________________: A certification approach for organizations planning to certify a large number of new construction buildings or interior spaces based on a common prototype, streamlining certification by focusing on similarities in building design and construction practices regardless of location.

5) ____________________________: A program designed for multiple buildings on a shared site under the control of a single entity. It allows project teams to document shared site-related credits once and apply them across all individual building certifications on the campus.

6) ____________________________: Committees that advise the LEED Steering Committee and consensus committees on specific technical aspects of LEED, such as materials or indoor air quality.

7) ____________________________: The overarching objectives that every LEED project aims to accomplish.

8) ____________________________: Credits being tested across all rating system types and credit categories and include credits proposed for the next version of LEED.

9) ____________________________: Provide additional options for meeting credit requirements for international projects that need to reference local or regional standards instead of United States–based standards.

ANSWER: MATCHING GAME

1) LEED Steering Committee (LSC)	4) LEED Volume Program	7) Impact Areas
2) LEED Project Administrator	5) LEED Campus Program	8) Pilot Credits
3) LEED Interpretation	6) Technical Advisory Groups (TAGs)	9) Alternative Compliance Paths (ACPs)

PRACTICE EXAM

To access the end-of-chapter practice exam online in our exam simulator, scan the QR code below with your laptop, phone, tablet, or another compatible device, then log in using your email address and password. If you have trouble scanning the QR code, visit www.projectific.com/account and log in using your email address and password. If you have not yet created a password, please refer to the *How to Access the Exam Simulator* section at the back of the book for instructions.

CHAPTER 4:

INTEGRATIVE PROCESS, PLANNING, AND ASSESSMENTS

This credit category will sound very familiar after having read the integrated process section in Chapter 1. Concepts such as systems thinking, feedback loops, life-cycle approach, and iterative process all feed directly into the prerequisites and credits in this credit category. In Chapter 1, we discussed how an integrated process brings together diverse stakeholders early in the project to set goals, evaluate alternatives, and make decisions that consider the building as a whole system rather than a collection of isolated parts. This credit category takes those principles and turns them into specific, measurable requirements.

The category contains three prerequisites and one credit as follows:

- Climate Resilience Assessment prerequisite
- Human Impact Assessment prerequisite
- Carbon Assessment prerequisite
- Integrative Design Process credit

Two concepts that are worth understanding before diving into the prerequisites and credits are resilience and responsiveness to community needs, which will be discussed next.

RESILIENCE

Buildings are designed to last for decades, but the climate conditions they will face over their lifetime are changing. Extreme weather events such as flooding, heat waves, wildfires, and severe storms are becoming more frequent and more intense, and most building codes have not caught up with these projected changes. This is where resilience comes in.

Resilience refers to a building's ability to prepare for, withstand, respond to, and recover from disruptive events. A resilient building is one that can continue to function and protect its occupants even when faced with unexpected conditions. Think of it as designing for what could happen, not just what has happened in the past.

The starting point for resilience planning is understanding what threats a project might face. A hazard is any event or condition that could cause harm—flooding, extreme heat, high winds, wildfires, and more. Not every project faces the same hazards; a coastal building in Florida deals with very different threats than a mountain facility in Colorado. A **hazard assessment** evaluates which threats are most relevant to the project by looking at climate projections alongside the specific conditions of the site.

Once the hazards are identified, the next step is understanding the risks they pose. While a hazard describes what could happen, **climate risk** refers to the potential adverse consequences of that hazard actually occurring. A **climate risk assessment** estimates the potential impact of those hazards on the building, its occupants, and the owner's investment. For example, the risk of a flood is not just the cost of repairing water damage—it also includes lost revenue from downtime, potential liability if occupants are harmed, and long-term issues like mold growth that can affect indoor air quality for years.

By addressing resilience early in the integrative process, project teams can evaluate whether the cost of protective measures is justified by the risks they mitigate.

RESPONSIVENESS TO COMMUNITY NEEDS AND SOCIAL EQUITY

Sustainability is not only about protecting the environment and saving energy. It is equally about the well-being of people—the communities that surround a building, the occupants who use it, and the workers who construct it and manufacture its materials.

Health outcomes are not determined by genetics alone. **Social determinants of health**—the conditions in which people are born, live, and work—play a significant role in shaping well-being. These include factors such as access to clean air, safe housing, green spaces, nutritious food, and quality education. Unfortunately, these conditions are not distributed equally. Low-income communities are far more likely to be located near sources of pollution such as industrial facilities or busy highways.

Green building can help address these inequities. **Environmental justice** is the principle that every person deserves equal protection from environmental and health hazards, equal access to decision-making, and equitable access to a healthy and sustainable environment. When project teams engage community members early in the planning process, they gain a deeper understanding of local needs and can design solutions that make a real difference. For instance, a residential building near an industrial zone might need enhanced air filtration to shield occupants from nearby emissions. An affordable housing project in a dense urban area might incorporate native landscaping and recreational areas designed for physical activity to ensure residents have access to nature and wellness opportunities despite limited private outdoor space.

These considerations do not stop at the project's boundaries. The people who physically build our projects also deserve attention. Without proper protections, construction workers can be exposed to extreme temperatures, hazardous chemicals, and dangerous working conditions. Project teams that plan for sustainable construction should ensure that safety practices and training are aligned with the project's broader social goals.

The same thinking applies to the global supply chain behind building materials. Every product installed in a building was manufactured, transported, and assembled by people—often in challenging conditions across multiple countries. Project teams can make a difference by choosing materials from manufacturers that demonstrate fair labor practices and by seeking certifications that verify worker health and safety protections throughout the supply chain.

INTEGRATIVE PROCESS, PLANNING, AND ASSESSMENTS PREREQUISITES AND CREDITS

CLIMATE RESILIENCE ASSESSMENT—PREREQUISITE

PREREQUISITE SUMMARY

Designing and developing buildings with climate resilience in mind can reduce climate-related damage and improve long-term performance and safety. In this context, a climate resilience assessment helps teams understand which site-specific hazards could threaten building safety, durability, and operations and supports selecting data-driven strategies.

This prerequisite requires project teams to complete a climate and natural hazard assessment that considers observed and projected hazards that could affect the project site and building operations, such as extreme heat, drought, flooding, hurricanes, wildfires, tornadoes, and other relevant threats. Teams must prioritize at least two hazards and analyze them using emissions scenarios from the Intergovernmental Panel on Climate Change's Shared Socioeconomic Pathways. The **Intergovernmental Panel on Climate Change** (**IPCC**) is the United Nations body for assessing the science related to climate change, and the **Shared Socioeconomic Pathways** (**SSPs**) are scenarios used in climate research to explore how global society, demographics, and economics might evolve over the 21st century, influencing greenhouse gas emissions and the severity of climate change.

As a result of this assessment, project teams can identify key vulnerabilities and use these findings to inform planning, design, and operations and maintenance to ensure long-term safety.

PREREQUISITE REQUIREMENTS

This prerequisite requires the project team to complete a climate and natural hazard assessment. The project team must identify the observed, projected, and future natural hazards that could affect the project site and the building operations. Hazards to consider include, but are not limited to, drought, extreme heat, extreme cold, flooding, hurricanes/high winds, hail, landslides, sea level rise and storm surge, tornadoes, tsunamis, wildfires and smoke, winter storms, and any other relevant local hazards.

After listing the hazards, the team must prioritize at least two priority hazards and show how the project will address them through design strategies.

For each priority hazard, the assessment must include:

- Intergovernmental Panel on Climate Change (IPCC) emissions scenario used, specifying the shared socioeconomic pathways
- Projected service life of the LEED project
- Hazard level
- Hazard risk rating
- Exposure, sensitivity, adaptive capacity, vulnerability, and overall risk levels
- Potential impact on the site and building function
- Potential impact on the site during construction

The findings of this assessment should inform planning, design, and operations and maintenance of the project.

KEY THINGS TO REMEMBER

- LEED requires all projects to complete a climate and natural hazard assessment and use the findings to inform the planning, design, and operations and maintenance of the project.
- LEED refers to the Intergovernmental Panel on Climate Change (IPCC) for the climate resilience assessment.

HUMAN IMPACT ASSESSMENT—PREREQUISITE

PREREQUISITE SUMMARY

This prerequisite requires project teams to evaluate the social, economic, and environmental context of the project's local community, workforce, and supply chain before development begins to ensure that the project aligns with community needs and promotes equitable outcomes. For this purpose, the prerequisite requires project teams to complete a human impact assessment by examining key characteristics unique to the project site and its surrounding community across four categories: demographics, local infrastructure and land use, human use and health impacts, and occupant experience. The findings should then be used to inform the planning, design, operations, and maintenance of the project and to help the team choose project-specific actions that respond to local needs. This process helps balance environmental goals with the health, safety, and welfare of the people impacted by the project, and it gives teams a structured way to identify disparities and incorporate strategies that support inclusivity, fair labor practices, and equitable access to opportunities.

PREREQUISITE REQUIREMENTS

Project teams should complete a human impact assessment by pulling relevant information from the categories below, as applicable:

- Demographics: Evaluate race and ethnicity, gender, age, income, employment rate, education levels, and household types.
- Local infrastructure and land use: Evaluate adjacent transportation and pedestrian infrastructure, adjacent diverse uses, relevant local sustainability goals, and applicable accessibility codes.
- Human use and health impacts: Evaluate housing affordability and availability, access to social services (e.g., healthcare, education, and social support networks), community safety and local community groups, and supply chain and construction workforce protections.
- Occupant experience: Evaluate the availability of natural daylight, quality of views, and the opportunity to provide operable windows. Evaluate air and water quality. Consider how surrounding buildings affect the indoor environment.
- Other, as applicable.

Project teams should use the findings of the human impact assessment to inform the project's planning, design, and operations and maintenance.

CARBON ASSESSMENT—PREREQUISITE

PREREQUISITE SUMMARY

This prerequisite helps project teams understand—and ultimately reduce—the project's long-term carbon footprint not just from operational energy use but also from construction-related emissions (embodied carbon) and refrigerant impacts. The idea is to give a clearer picture of where emissions are coming from over time so project teams and owners can focus their decarbonization efforts where they matter most.

USGBC will generate a 25-year carbon assessment using information already submitted under specific prerequisites that will be mentioned shortly so that project teams will be able to make more informed carbon-related decisions.

PREREQUISITE REQUIREMENTS

USGBC will produce a 25-year carbon assessment for the project using data the team already submits under the Operational Carbon Projection and Decarbonization Plan prerequisite, Fundamental Refrigerant Management prerequisite, and Quantify and Assess Embodied Carbon prerequisite, and it may also include the Transportation Demand Management credit if the project pursues that credit. The assessment reports the project's annual emissions from operations, refrigerants, and embodied carbon (and transportation, if applicable) in detail for 25 years also showing each source's emissions as a percentage of the total.

Project teams do not need to make any calculations or submit any documentation to USGBC to earn this prerequisite.

KEY THINGS TO REMEMBER

- Operational energy use is not the only major source that contributes to greenhouse gas emissions. Emissions from construction (embodied carbon) and refrigerants are also considerable and must be evaluated by the project team.

INTEGRATIVE DESIGN PROCESS—CREDIT

- 1 Point (New Construction)

CREDIT SUMMARY

This credit will sound very familiar after reading the Integrative Process section. It awards one point to projects that implement an integrative design process by meeting the credit requirements.

As mentioned previously, collaboration among team members across disciplines is essential to achieving successful project outcomes. Each building system should be addressed from the predesign phase through rigorous questioning and a deliberate challenge of typical project assumptions.

The first step in earning this credit is forming an integrated team that brings together diverse expertise and viewpoints, including but not limited to the owner, architect, engineers (e.g., structural, mechanical, and electrical), contractor, future building users, and community representatives. The integrated team should collaboratively evaluate multiple facets of the project across all phases—from early design through construction and operations—by setting goals, developing strategies, and optimizing performance, feasibility, and costs.

During predesign or early design, the credit requires the team to conduct a design charrette with the owner (or the owner's representative) and participants representing at least four key perspectives (e.g., general contractor, lead architect, key engineering disciplines, future building users, and community representatives). **Charrettes** are intensive, multidisciplinary workshops held at the beginning of a project and at key milestones. Charrettes allow brainstorming and collaboration among different disciplines and experiences. As a result of the collaboration of diverse stakeholders, charrettes also enable the project team to identify innovative approaches to design and construction and to establish new or refined project goals.

As a result of this integrative design process, LEED requires projects to define a set of specific, measurable goals that address the LEED v5 impact areas. These goals should be incorporated into the owner's project requirements and tracked through design via the basis of design and other project documents. As its name implies, the **owner's project requirements** (**OPR**) document the owner's vision and performance requirements, including the LEED objectives the project needs to accomplish. On the other hand, the **basis of design** (**BOD**) describes the information necessary to accomplish the owner's project requirements, including system requirements, design criteria, standards, guidelines, etc., developed by the architect/engineer.

To become familiar with the integrative process, project teams are encouraged to review the **ANSI Consensus National Standard Guide 2.0 for the Design and Construction of Sustainable Buildings and Communities.**

CREDIT REQUIREMENTS

Project teams should begin an integrative design process during predesign and continue it through early occupancy to identify and implement strategies that create synergies across building systems through the following initiatives:

- Integrated team: Form and regularly meet with an interdisciplinary team that brings different expertise and viewpoints.
- Design charette: In predesign or early design, run a charette with the owner (or owner's representative) and participants representing at least four key perspectives.
- LEED goal setting: Collaboratively set specific, measurable project goals that align with the LEED v5 impact areas of decarbonization, quality of life, and ecological conservation and restoration, and incorporate them into the owner's project requirements.

KEY THINGS TO REMEMBER

- As a result of the integrative design process, LEED requires projects to define a set of specific, measurable goals that address the LEED v5 impact areas. These goals should be incorporated into the owner's project requirements and tracked through design via the basis of design and other project documents.
- To become familiar with the integrative process, project teams are encouraged to review the ANSI Consensus National Standard Guide 2.0 for the Design and Construction of Sustainable Buildings and Communities.

STRATEGIES TO ADDRESS INTEGRATIVE PROCESS, PLANNING, AND ASSESSMENTS CREDIT CATEGORY

- Implement an integrative design process that brings together diverse stakeholders and specialists from the earliest stages of the project.
- Collaboratively set specific, measurable project goals that align with the LEED v5 impact areas of decarbonization, quality of life, and ecological conservation and restoration, and incorporate them into the owner's project requirements.
- Conduct a carbon assessment early in the design process to understand the project's total carbon footprint.
- Analyze the embodied carbon of building materials and construction activities.
- Evaluate long-term carbon emissions associated with the building's energy consumption.
- Estimate the emissions generated by occupants and visitors commuting to and from the project and explore strategies to reduce transportation-related carbon.
- Conduct a climate resilience assessment to identify natural hazards and climate-related threats relevant to the project's location.
- Perform a climate risk assessment to estimate the potential consequences of identified hazards.
- Develop resilience strategies that allow the building to prepare for, withstand, respond to, and recover from disruptive events.
- Engage community stakeholders early in the planning process to ensure the project reflects local needs and priorities.
- Conduct a human impact assessment to evaluate how the project affects surrounding communities and populations.
- Study the demographics of the surrounding area and identify any vulnerable populations, including fenceline communities, that may be affected by the project.
- Evaluate race and ethnicity, gender, age, income, employment rate, education levels, and household types in the project's surrounding area.
- Review local infrastructure, land-use policies, and zoning conditions that may influence the project's social and environmental impact on the community.
- Evaluate adjacent transportation and pedestrian infrastructure, adjacent diverse uses, relevant local sustainability goals, and applicable accessibility codes.
- Evaluate the availability of natural daylight, quality of views, and the opportunity to provide operable windows.
- Evaluate air and water quality in the project's surrounding area.
- Use the findings of the human impact assessment to inform the project's planning, design, and operations and maintenance.
- Address social determinants of health by designing spaces that improve access to clean air, safe environments, green spaces, and community resources.
- Ensure appropriate workforce protections are in place so that construction crews are not exposed to unsafe working conditions.
- Select building materials from manufacturers that demonstrate fair labor practices and seek certifications that verify worker health and safety protections throughout the supply chain.
- Consider the environmental effects of ongoing purchases related to building operations and maintenance, such as cleaning products, replacement materials, and landscaping supplies.

END OF CHAPTER EXERCISES

EXERCISE: MATCHING GAME

Match the following terms to their definition.

Terms		
Resilience	Basis of Design (BOD)	Environmental Justice
Social Determinants of Health	Owner's Project Requirements (OPR)	

1) ___________________________: The principle that every person deserves equal protection from environmental and health hazards, equal access to decision-making, and equitable access to a healthy and sustainable environment.

2) ___________________________: A building's ability to prepare for, withstand, respond to, and recover from disruptive events.

3) ___________________________: Describes the information necessary to accomplish the owner's project requirements, including system requirements, design criteria, standards, guidelines, etc., developed by the architect/engineer.

4) ___________________________: The conditions in which people are born, live, and work that play a significant role in shaping well-being, including factors such as access to clean air, safe housing, green spaces, nutritious food, and quality education.

5) ___________________________: A document that outlines the owner's vision and performance requirements for a project, including the LEED objectives the project needs to accomplish.

ANSWER: MATCHING GAME

1) Environmental Justice	3) Basis of Design (BOD)	5) Owner's Project Requirements (OPR)
2) Resilience	4) Social Determinants of Health	

PRACTICE EXAM

To access the end-of-chapter practice exam online in our exam simulator, scan the QR code below with your laptop, phone, tablet, or another compatible device, then log in using your email address and password. If you have trouble scanning the QR code, visit www.projectific.com/account and log in using your email address and password. If you have not yet created a password, please refer to the *How to Access the Exam Simulator* section at the back of the book for instructions.

CHAPTER 5:

LOCATION AND TRANSPORTATION

This chapter will discuss the importance of location and the transportation features of the building and their effects on the green building design. After discussing the LT credit category, we will go over all the LT credits individually to show how these features are used in LEED certification for a LEED BD+C: New Construction and Major Renovations project. (The LT credit category does not have any prerequisites.)

The Location and Transportation credit category will be discussed under four major sections:

- Location
- Transportation
- Neighborhood Pattern and Design
- Equitable Development and Community Impact

LOCATION

The location of a green building project should first promote **smart growth,** which refers to the approach that protects undeveloped lands and contributes to developments in projects near jobs, schools, shops, and other destination points with diverse uses.

An example may be a residential project that is located very close to downtown, which also contains several public transportation options.

However, a residential project in a suburban area that requires driving many miles to the downtown area for people to commute would not contribute to smart growth, and that project would hence support suburban sprawl.

There are also other factors to consider when choosing a proper project location. Starting from smart growth, the book will go over all the factors that will contribute to a green building's location.

SMART GROWTH

Suburban sprawl has lots of consequences to the environment. First, the construction of a building on undeveloped land will destroy the habitat and wildlife in that location. Second, suburban sprawl will result in car dependency, which will further damage the environment with the greenhouse gases created. In addition, after people start moving to suburban areas, the cities will need to provide infrastructure to the suburban areas, which will create additional consequences for the environment.

Principles of smart growth are:

- Protect undeveloped land (undeveloped land is also called **greenfield**)
- Reuse/restore previously developed sites
- Reduce automobile use and promote public transportation
- Develop efficient rainwater management
- Reduce the heat island effect (will be discussed in the next chapter)
- Reduce lighting pollution
- Provide stewardship of nature and the site's surroundings

According to the U.S. Environmental Protection Agency (EPA), in order to promote smart growth and discourage suburban sprawl, the following strategies should be applied[1]:

- Mix land uses (residential, commercial, retail, etc. all on the same land)
- Take advantage of compact building design
- Create a range of housing opportunities and choices
- Create walkable neighborhoods
- Foster distinctive, attractive communities with a strong sense of place
- Preserve open space, farmland, natural beauty, and critical environmental areas
- Strengthen and direct development toward existing communities
- Provide a variety of transportation choices
- Make development decisions predictable, fair, and cost effective
- Encourage community and stakeholder collaboration in development decisions

To promote smart growth, some municipalities offer increased **floor-to-area ratios (FAR)** in urban areas. The floor-to-area ratio is also a very important term for the exam, and it is calculated by dividing the total square feet of a building by the total square feet of the lot of the building. For example, 10,000 square feet of land that has a FAR of 2 can allow the construction of a 20,000-square-foot building. If the building has two stories, each story can contain 10,000 square feet of space, which in this case, the building would cover the whole lot since the lot also measures 10,000 square feet. (These types of building, which are built on the entire lot, are called zero-lot-line projects.)

Another 10,000 square feet of land that has a FAR of 1 would allow construction of a 10,000-square-foot building. Again, if the building has two stories, and each story contains 5,000 square feet of space, in which case the remaining 5,000 square feet of the lot could contain parking or landscape. Hence, a higher FAR allows more building space and a denser development.

PROTECTING THE HABITAT BY CHOOSING THE RIGHT LOCATION

To protect the habitat, green buildings should not be developed inside **sensitive lands**; instead, **infill sites** should be preferred. Infill sites, or infill developments, are sites that were either previously developed or were already being used for other purposes in urban areas. An example of an infill site could be a project location inside the downtown of a city that was used as a parking lot. Or it could be the site of a demolished building. Since infill sites have existing infrastructure and also have public transportation options, locating the project at an infill site would not create the negative consequences of suburban sprawl.

Below are the types of sensitive lands that should be avoided for development:

- Prime farmland
- Sites close to wetlands and water bodies (lakes, rivers, etc.)
- Public parkland
- Areas below floodplain
- Areas that are a part of the habitat or endangered species

Locating the project on brownfield sites and remediating them before the start of construction can be another option, which can also gain extra points in the LEED certification.

Brownfield sites are previously developed sites that were contaminated with waste or pollution. A site that has an abandoned building with unknown contamination can also be classified as a brownfield site. Once the contamination is cleaned up, these sites can be an excellent option for project development, and the federal, state, or municipal government can offer tax incentives on the property to support development.

Brownfield Site *(Photo courtesy of Petr Vilgus)*

TRANSPORTATION

Another key component of LEED is transportation, which accounts for 50% of a project's total greenhouse gas emissions and 33% of greenhouse gas emissions in the United States, according to the U.S. Energy Information Administration.[2] The transportation sector is also responsible for about one-quarter of energy-related greenhouse gas emissions globally.[3]

A project should reduce the consequences of transportation by ensuring access to alternative modes of transportation to reduce single-occupancy vehicle use; encouraging walking and bicycling; and actively supporting electric vehicle adoption through the installation of electric vehicle supply equipment (EVSE). If the project is located in a high-density area or infill site, building occupants can walk to diverse uses and find different options for using public transit. Note that human-powered mobility such as biking, walking, or wheeling is referred to as **active transportation**.

Project sites without access to public transportation can focus on local connectivity and may promote the use of electric vehicles for commuting, provide incentives for carpooling, or develop diverse uses that allow workers to walk to basic services.

Utilizing a **compressed workweek** can be another strategy to reduce fossil fuel usage. A compressed workweek typically occurs when a company allows its employees to work more hours during working days in exchange for a day off on a weekday. Alternatively, companies could utilize **telecommuting** for a part of the workweek so that employers could work from their homes without commuting to the office. Carpooling can also be rewarded, and parking rates could be made more expensive to discourage individual vehicle use.

LEED also promotes limiting the number of parking spaces and/or charging a parking fee equal to or greater than the local market rate for public or private parking to discourage individual vehicle use. This topic is discussed further under the Transportation Demand Management credit.

If the development location is close to a bicycle network, providing secure bicycle storage, shower rooms, and bike repair stations helps building occupants commute by bicycle more easily and incorporate physical activity into their daily routines.

NEIGHBORHOOD PATTERN AND DESIGN

To discourage suburban sprawl, we should pay special attention to the neighborhood pattern and design. A healthy neighborhood should contain wide sidewalks, benches, and bicycle networks. Business centers, retail services, educational facilities, and other diverse uses should be close enough to minimize travel. Public transportation options should be easily accessible. Street layouts should allow for easy connectivity, and if community gardens, farmers markets, and agricultural programs are established, a neighborhood could also be able to support access to sustainable food.

The aforementioned strategies are also a part of **compact development strategies**, also called high-density development strategies, which promote location-efficient neighborhoods with high development density and reduce greenhouse gas emissions. **Development density** is a measure of the total building floor area or total number of dwelling units on a unit of land relative to the total buildable area on that land. **Location efficiency** refers to how well a project's surroundings reduce the need for driving by providing easy access to transit, services, and amenities within walking or cycling distance. Manhattan, New York is a well-known example of location-efficient, **high-density development**, while a rural subdivision with large single-family lots and wide spacing between homes is an example of **low-density development**.

EQUITABLE DEVELOPMENT AND COMMUNITY IMPACT

When choosing the location of a project, we also need to be sensitive to the people who already live there. While dense, well-connected neighborhoods can offer many advantages, new development should not come at the expense of affordability or force existing residents out of their communities. For this reason, projects should be planned in a way that supports inclusion and equity from the outset.

Project teams should ensure that the project is inclusive and equitable. In this regard, they may set aside affordable units for people whose incomes are lower than the area median.

Projects can further support community well-being by incorporating spaces and services that respond to local needs. To do so, project teams should engage with community members early in the process and continue that engagement throughout planning and design. Working with local nonprofits and public agencies can help teams better understand who the project should serve and what resources are most needed.

Certain project types, especially schools, can also play a broader community role. Because schools are often centrally located and have spaces that remain unused outside regular hours, they can support the surrounding neighborhood by opening amenities such as gymnasiums or auditoriums for public use or by providing dedicated space for services such as health clinics or social support agencies.

In addition to supporting residents, projects should also consider how construction can create equitable economic opportunities. Construction jobs can provide stable employment, benefits, and higher wages for workers without requiring a college degree. Project teams and contractors can extend these benefits by hiring workers from the surrounding area, offering apprenticeships and skills training, and creating pathways for long-term economic advancement. Support can also go beyond the jobsite through programs such as financial literacy training or entrepreneurship education.

STRATEGIES TO ADDRESS LOCATION AND TRANSPORTATION

- Locate the project in location-efficient, high-density development areas.
- Reuse or renovate an existing building, or develop an infill site.
- Protect habitat by avoiding development on sensitive lands such as prime farmlands, sites close to wetlands and water bodies, public parklands, areas below the floodplain, and areas that are part of the habitat of endangered species.
- Choose brownfields for project development by first remediating the entire site.
- Locate the project near existing infrastructure.
- Choose a project location within walking distance of diverse uses. Examples of diverse uses include banks, restaurants, schools, retail shops, and more.
- Promote alternative transportation, which refers to the use of modes of transportation other than single-occupancy vehicles.
- Locate the project near public transit.
- Limit the number of parking spaces in the project and charge a parking fee equal to or greater than the local market rate to discourage individual vehicle use.
- Promote bicycling by providing secure bicycle storage, shower rooms, and bike repair stations.
- Encourage car-share programs.
- Encourage carpooling.
- Actively support electric vehicle adoption by installing electric vehicle supply equipment (EVSE) as part of the project.
- Consider providing affordable housing units.
- Support community well-being by incorporating spaces and services that respond to local needs.
- Engage with community members early to better understand local priorities and needs.
- In school projects, consider allowing after-hours public use of certain amenities such as gymnasiums or auditoriums.
- Dedicate space within the project for community-serving functions such as health clinics, social support services, or other essential services.
- Support equitable economic opportunities during construction by hiring workers from the surrounding area whenever possible.
- Provide apprenticeships, job-skills training, and other workforce development opportunities during the construction process.
- Enhance long-term economic opportunity by supporting construction workers through programs such as financial literacy training or entrepreneurship education.

LOCATION AND TRANSPORTATION CREDITS

SENSITIVE LAND PROTECTION—CREDIT

- 1 Point (New Construction)

CREDIT SUMMARY

This credit is designed to prohibit site development in ecologically sensitive areas. Ecologically sensitive areas provide a variety of benefits not only for human health but also for the environment. By locating the building's footprint on a previously developed area, ecologically sensitive areas are preserved.

If the project is located on previously developed land, it will earn the credit because the land has already lost its ecological value. However, if the land, or a portion of the land, has not been previously developed, then no portion of the development footprint should be located in any sensitive areas, including prime farmland, floodplain, notable habitat, water bodies, or wetlands.

CREDIT REQUIREMENTS

Option 1: Previously Developed Sites (1 Point)

Locate the development footprint on previously developed land.

OR

Option 2: Previously Undeveloped Sites (1 Point)

If the site has not been previously developed, the development footprint should avoid the following types of sensitive land:

- Prime farmland: Land classified as prime farmland, unique farmland, or farmland of statewide or local importance, as defined by the **U.S. Department of Agriculture** in the U.S. Code of Federal Regulations (Title 7, Volume 6, Parts 400–699, Section 657.5) (or a local equivalent for projects outside the U.S.), and identified in a **Natural Resources Conservation Service** (**NRCS**) soil survey (or local equivalent for projects outside the U.S.).
- Floodplains: Areas shown as floodplains on **Federal Emergency Management Agency** (**FEMA**) flood hazard maps or maps from local authorities. If no maps are available, the project team must confirm the site is not in an area with a 1% or greater annual chance of flooding in any year.

- Notable habitat: Land identified as habitat for any of the following.
  - Species listed as threatened or endangered under the **U.S. Endangered Species Act** or a state endangered species act.
  - Species or ecological communities classified by NatureServe as GH (possibly extinct), G1 (critically imperiled), or G2 (imperiled).
  - Species listed as threatened or endangered under equivalent local standards (for projects outside the U.S.) that are not covered by NatureServe data, such as the **International Union for Conservation of Nature (IUCN) of Nature's Red List**.
- Water bodies: No development is permitted within 100 feet (30 meters) of a water body, except for minor improvements.
- Wetlands: No development is permitted within 50 feet (15 meters) of a wetland, except for minor improvements.
- Steep slopes: Protect at least 40% of the site's steep-slope areas (if present) from any development or construction activity.

KEY THINGS TO REMEMBER

- The U.S. Department of Agriculture defines prime farmland.
- The Federal Emergency Management Agency (FEMA) designates flood hazard areas.
- The U.S. Endangered Species Act defines endangered and threatened species.
- NatureServe defines and assigns conservation status ranks for species and ecosystems.
- The International Union for Conservation of Nature (IUCN) Red List identifies endangered and threatened species.

EQUITABLE DEVELOPMENT—CREDIT

- 1–2 Points (New Construction)

CREDIT SUMMARY

This credit rewards projects that make smarter choices about where they build, who benefits from the project's construction work, and how the project supports low-income people—not only through what gets built, but also through how the project team hires, trains, and contributes to the surrounding community.

The credit has three options. The first option focuses on priority sites, encouraging development on brownfields that need to be remediated or in historic districts that support responsible growth. As discussed, brownfield sites are previously developed sites that were contaminated with waste or pollution. Even an abandoned building with unknown contamination can fall into this category. Once the contamination is properly addressed, these sites can become excellent locations for new development, and in many cases, governments may offer incentives, such as tax benefits, to support redevelopment.

The second option focuses on proximity to housing and jobs, either by hiring locally for the project's construction work or by providing affordable housing in residential and mixed-use projects.

The third option focuses on equitable construction by investing in construction workers through on-the-job training programs or life-skills training.

CREDIT REQUIREMENTS

Option 1: Priority Sites (1–2 Points)

Under this option, project teams have two paths. In Path 1, the project must be located on a brownfield site where remediation is required by the authority having jurisdiction (local, state, or national). If the project team is remediating voluntarily, they must still obtain confirmation from the same authority verifying that the site qualifies as a brownfield. In either case, the cleanup must be completed to the satisfaction of the relevant authority.

In Path 2, the project should be located in a historic district that has been identified by the local government, typically as part of a growth management plan or policy.

AND/OR

Option 2: Housing and Jobs Proximity (1–2 Points)

Under this option, project teams have two paths. In Path 1, the focus is supporting the local economy through local hiring. Project teams must employ individuals who live within the project site's administrative district for at least 15% of the construction jobs created by the LEED project.

Path 2 focuses on affordable housing and is available for residential or mixed-use projects. This path requires that a portion of new dwelling units be provided as affordable rentals and/or for-sale homes, priced for households earning below the Area Median Income (AMI).

OR

Option 3: Equitable Construction (2 Points)

This option focuses on supporting construction workers on the project by providing access to workforce development training. Project teams earn the points by choosing one of two training approaches.

The first approach is job-related skills training, which is achieved through on-the-job training in a Department of Labor registered apprenticeship program (or a local equivalent outside the U.S.). To comply, at least 15% of the total project construction hours must be performed by workers enrolled in these registered apprenticeship programs.

The second approach is life-skills training, where an organization or government entity provides on-site courses for construction workers on topics such as financial literacy, debt management, first-time home buying, or entrepreneurship. Under this approach, the project must schedule one course per month for the entire duration of construction.

COMPACT AND CONNECTED DEVELOPMENT—CREDIT

- 1–6 Points (New Construction)

CREDIT SUMMARY

A project located within existing built density and/or within walkable distance of diverse uses such as banks, restaurants, and supermarkets has great benefits for both project users and the environment, as it contributes to compact development. Increasing residential and nonresidential densities in an area will reduce vehicle use and, consequently, reduce air pollution as well as greenhouse gas emissions. Car accidents will also be less likely to occur.

This credit awards points for projects that are located within surrounding density and/or within walking distance of diverse uses. If the project satisfies both criteria, then the total points from both options will be awarded.

Locating a project within walking distance of quality transit also provides significant benefits because it reduces single-occupancy vehicle use and associated air pollution, including greenhouse gas emissions. The environmental impacts of public transportation are generally far lower than those of single-occupancy vehicles. In addition, access to public transit can be especially helpful for people who do not own a car and for older adults who cannot drive. This credit also awards points to projects that have access to public transit or that provide project-sponsored transit service between the project site and external destinations.

CREDIT REQUIREMENTS

Option 1: Surrounding Density (1–2 Points)

Under this option, the project must be located on a site where the existing surrounding density within a 0.25-mile (400-meter) offset of the project boundary meets LEED's thresholds. Projects in areas with higher surrounding density can earn more points.

AND/OR

Option 2: Access to Transit (1–4 Points)

Under this option, project teams can follow one of two paths. In Path 1, the project must place a functional building entry within walking distance of existing or planned public transit. The entry must be within 0.25 miles (400 meters) of bus, streetcar, or informal transit stops, or within 0.5 miles (800 meters) of bus rapid transit, passenger rail stations, or commuter ferry terminals. The transit service at those stops must meet LEED's minimum weekday and weekend trip requirements, and projects with more frequent service earn more points.

In Path 2, the project should commit to providing project-sponsored transit service (such as shuttles or vans) that meets the minimum daily service levels defined by LEED. The transit service must provide transportation between the project site and external destinations such as residential areas and public transportation stations. This service should be guaranteed for at least three years after the certificate of occupancy, and should include at least one accessible transit stop shelter within 0.25 miles (400 meters) of a functional entry.

AND/OR

Option 3: Walkable Location (1–3 Points)

This option rewards projects that are located where people can walk to daily needs. Project teams can show compliance in two ways.

In the first route, the project site must meet the required Walk Score® range. **Walk Score** measures the walkability of a location primarily by looking at walking routes to nearby amenities. It is a number between 0 and 100. A higher Walk Score indicates more amenities and services within a short walking distance of the project.

In the second route, any functional entry of the building must be within 0.5 miles (800 meters) walking distance of a certain number of publicly available uses defined by LEED.

The following table shows the eligible use types and categories.

Category	Use type
Food retail	Supermarket
	Grocery with produce section
Community-serving retail	Convenience store
	Farmers market
	Pharmacy
	Hardware store
	Other retail
Services	Bank
	Family entertainment venue (theater)
	Gym, health club, exercise studio
	Laundry, dry cleaner
	Hair care
	Restaurant, cafe, diner
Civic and community facilities	Adult or senior care (licensed)
	Child care (licensed)
	Community or recreation center
	Cultural arts facility (e.g., museum, performing arts)
	Education facility (e.g., K-12 school, university)
	Education center, vocational school, community college
	Government office that serves public on-site
	Medical clinic or office that treats patients
	Place of worship
	Police or fire station
	Post Office
	Public library
	Public park
	Social services center
Community anchor uses	Commercial office (100 or more full-time equivalent jobs)

Note that no more than two uses in each use type may be counted. For example, if there are four restaurants within walking distance, only two may be counted. Additionally, the counted uses should represent at least three of the five use categories.

KEY THINGS TO REMEMBER

- To measure existing surrounding density of a project, LEED considers the 0.25-mile (400-meter) offset of the project boundary.
- For all LEED credits that require walking distance calculations, walkability should be confirmed using paths that provide safe and comfortable environments for pedestrians and are part of a continuous network of sidewalks. Any walking route that does not fulfill these requirements cannot be documented as being within walking distance.
- In LEED, 0.25 miles (400 meters) is the walking distance to bus, streetcar, or informal transit stops, and 0.5 miles (800 meters) is the walking distance to bus rapid transit stops, rail stations, and commuter ferry terminals.
- LEED refers to Walk Score to assess the walkability of the project's location to amenities and services.
- For a project to be considered walkable to access publicly available uses, LEED requires a walking distance of 0.5 miles (800 meters) from the project, which corresponds to a 10-minute walking time.
- LEED groups "publicly available uses" into five categories—food retail, community-serving retail, services, civic and community facilities, and community anchor uses.

TRANSPORTATION DEMAND MANAGEMENT—CREDIT

- 1–4 Points (New Construction)

CREDIT SUMMARY

This credit aims to reduce the pollution associated with daily automobile travel by building occupants. The credit starts by requiring all project teams to conduct a transportation demand assessment, where they estimate annual vehicle miles traveled (VMT), calculate a baseline carbon emissions case, evaluate low-carbon transportation options, calculate a proposed emissions case, and then quantify the total emissions reduction. If the project is already part of a government-mandated transportation demand management program, this assessment requirement is automatically satisfied. A **transportation demand management (TDM)** program is a set of rules and strategies designed to reduce single-occupancy vehicle trips. In practice, it might include measures like public transit incentives, carpool programs, employer commuter benefits, limits on parking supply, and policies that make walking, biking, and public transit more attractive.

Parking spaces create a considerable amount of harm to the environment. First, the surfaces of parking lots trap heat (especially dark-colored surfaces); this results in heat island effects, which increase ambient air temperatures in urban areas. We have all seen that, in hot climates, asphalt parking lots release heat just like a hair dryer blowing hot air. However, if a greenfield were there, this effect wouldn't occur. In addition, because of the impervious surfaces of parking lots, rainwater runoff occurs, which can overwhelm municipal stormwater systems and carry contaminants into waterways. For those reasons, Option 1 of this credit provides points to projects that limit the number of parking spaces and/or charge a parking fee that is equal to or greater than the local market rate for public or private parking.

Option 2 of this credit, on the other hand, promotes bicycling, which also has great benefits for human health. Under this option, project teams can earn points through several strategies, such as locating or designing the project so that a functional entry and/or bicycle storage is within 600 feet (180 meters) walking or bicycling distance of a qualifying bicycle network, providing short-term and long-term bicycle storage, offering showers and changing facilities, and even installing a bicycle repair station.

CREDIT REQUIREMENTS

Transportation Demand Assessment:

To satisfy the Transportation Demand Assessment requirement, project teams should:

- Estimate the annual VMT
- Estimate the annual baseline carbon emissions case
- Evaluate low-carbon transportation options
- Estimate the annual proposed carbon emissions case

- Calculate the total carbon reduction between the baseline and proposed cases

Projects that participate in a local or regional, government-mandated transportation demand management (TDM) program automatically satisfy this requirement.

AND

Option 1: Parking (1–3 Points)

Under this option, project teams have two paths.

In the first path, project teams should reduce parking capacity using base ratios from the **Institute of Transportation Engineers Parking Generation Manual**, **6th edition**, or a comparable resource. Points are based on the percent reduction, and projects that make a 30% reduction earn 1 point while projects making a 100% reduction (no parking) earn 3 points.

In the second path, projects can earn 2 points by charging a daily, monthly, or annual parking fee that is equal to or greater than the local market rate for public or private parking.

AND/OR

Option 2: Active Travel Facilities (1–3 Points)

Under this option, project teams have three paths.

In the first path, the project must be located or designed so that a functional entry and/or bicycle storage is within 600 feet (180 meters) walking or bicycling distance of a qualifying bicycle network. The bicycle network must be continuous for at least 3 miles (4,800 meters) and consist of bicycle paths, bicycle lanes, multiuse trails, or streets with a maximum speed limit of 25 mph (40 kph). Planned bicycle facilities may also count if they are fully funded by the certificate of occupancy date and scheduled for completion within three years.

Additionally, the project should contain short-term bicycle storage within 600 feet (180 meters) of any main entrance (minimum four spaces per building) and long-term storage within 300 feet (90 meters) of any functional entry (minimum four spaces per building).

In the second path, the project should include showers and changing facilities for 1% of regular building occupants. Off-site facilities may be used if they serve all occupants and are within 0.25 miles (400 meters) walking distance.

In the third path, the project should include a permanent, secure bicycle repair station with a complete set of tools and an air pump, located in the long-term bicycle storage area.

KEY THINGS TO REMEMBER

- To reduce the pollution associated with daily automobile travel by building occupants, LEED encourages projects to complete a transportation demand assessment and reduce driving by limiting parking spaces and/or charging market-rate parking fees. LEED also supports bicycling and awards projects that have close access to a qualifying bicycle network and that provide bicycle storage, showers, changing facilities, and a bicycle repair station.
- LEED refers to the Institute of Transportation Engineers Parking Generation Manual for base parking capacity ratios.

ELECTRIC VEHICLES—CREDIT

- 1–2 Points (New Construction)

CREDIT SUMMARY

In the United States, transportation accounts for 27% of total greenhouse gas emissions from the combustion of petroleum-based fuels. The use of petroleum-based fuels is a serious contributor to climate change and a major source of air pollution. For that reason, this credit aims to increase electric vehicle adoption to help reduce the negative effects of petroleum-based vehicle use.

Option 1 of this credit rewards projects that install electric vehicle supply equipment (EVSE). Option 2, on the other hand, recognizes projects that make parking spaces EV-ready, meaning the electrical capacity, conduit, and wiring are already in place so chargers can be added later easily.

CREDIT REQUIREMENTS

Option 1: Electric Vehicle Supply Equipment (1–2 Points)

This option awards points based on how many parking spaces have EVSE. For commercial projects, project teams earn 1 point by installing EVSE for at least 5% of total parking spaces (or at least 2 spaces, whichever is greater), and 2 points by reaching 10% (or at least 4 spaces, whichever is greater). For residential projects, project teams earn 1 point at 10% (or at least 5 spaces, whichever is greater), and 2 points at 15% (or at least 10 spaces, whichever is greater).

Installed EVSE must meet the following criteria:

- Provide Level 2 or Level 3 charging, following the manufacturer's instructions and **National Electrical Code (NFPA 70)**.
- Deliver 208–240 volts or higher to each required charging space.
- Meet the connected functionality criteria for **ENERGY STAR-certified EVSE**.
- Provide at least one accessible charging space that is 9 ft (2.5 m) wide with a 5 ft (1.5 m) access aisle, plus accessibility features for users with mobility, ambulatory, and visual limitations.

AND/OR

Option 2: Electric Vehicle Readiness (1 Point)

Provide electric vehicle-ready infrastructure to ensure future EVSE readiness. Commercial projects can earn 1 point by providing electric vehicle-ready infrastructure for at least 10% of all parking spaces or at least 10 spaces, whichever is greater. Residential projects can earn 1 point by providing EV-ready infrastructure for at least 20% of all parking spaces or at least 20 spaces, whichever is greater. Each EV-ready space must include a full-circuit installation, including 208–240 volts, 40-amp panel capacity, and a conduit (raceway) with wiring that terminates in a junction box or charging outlet.

KEY THINGS TO REMEMBER

- For electric vehicle supply equipment, LEED refers to the National Electric Code (NFPA 70) and ENERGY STAR.

END OF CHAPTER EXERCISES

EXERCISE: MATCHING GAME

Match the following terms to their definition.

Terms		
Smart Growth	Location Efficiency	Transportation Demand Management (TDM)
Brownfield Sites	Development Density	
Active Transportation	Walk Score	

1) ______________________: A measure of how well a project's surroundings reduce the need for driving by providing easy access to transit, services, and amenities within walking or cycling distance.

2) ______________________: A measure of the total building floor area or total number of dwelling units on a unit of land relative to the total buildable area on that land.

3) ______________________: A measure of a location's walkability on a scale of 0 to 100, based on walking routes to nearby amenities. A higher score indicates more amenities and services within a short walking distance.

4) ______________________: A neighborhood development approach that protects undeveloped lands and contributes to project development in locations near jobs, schools, shops, and other diverse uses.

5) ______________________: Previously developed sites that were contaminated with waste or pollution.

6) ______________________: Human-powered mobility such as biking, walking, or wheeling.

7) ______________________: A set of rules and strategies designed to reduce single-occupancy vehicle trips, including measures such as public transit incentives, carpool programs, employer commuter benefits, limits on parking supply, and policies that encourage walking, biking, and public transit.

ANSWER: MATCHING GAME

1) Location Efficiency	4) Smart Growth	7) Transportation Demand Management (TDM)
2) Development Density	5) Brownfield Sites	
3) Walk Score	6) Active Transportation	

PRACTICE EXAM

To access the end-of-chapter practice exam online in our exam simulator, scan the QR code below with your laptop, phone, tablet, or another compatible device, then log in using your email address and password. If you have trouble scanning the QR code, visit www.projectific.com/account and log in using your email address and password. If you have not yet created a password, please refer to the *How to Access the Exam Simulator* section at the back of the book for instructions.

CHAPTER 6:
SUSTAINABLE SITES

This chapter will discuss both the importance of sustainable sites and their effects on the green building design. After discussing the SS credit category, the book will go over all the SS prerequisites and credits individually to show how these features are used in LEED certification for a LEED BD+C: New Construction and Major Renovations project.

The Sustainable Sites credit category will be discussed under six major sections:

- Site assessment
- Site design
- Minimizing construction impacts
- Rainwater management
- Heat island effect
- Site management

SITE ASSESSMENT

As an important aspect of the integrated process of creating a sustainable site, a site assessment first needs to be conducted in order to have the relevant information that will be necessary for the site design.

The project team can conduct an assessment evaluating the following site features:

- **Topography:** Contour mapping, slope stability risks, unique topographic features
- **Hydrology:** Floor hazard areas, delineated wetlands, lakes, streams, shorelines, rainwater collection, and reuse
- **Climate:** Solar exposure, seasonal sun angles, winds, precipitation values
- **Vegetation:** Primary vegetation types, significant tree mapping, threatened or endangered species, habitat
- **Soils:** Prime farmland, healthy soils
- **Human use:** Views, proximity to transportation, proximity to diverse uses, adjacent properties
- **Human health effects:** The degree of air pollution on-site, the level of contamination present on-site, proximity to vulnerable populations

SITE DESIGN

With the completion of the site assessment, the project team will have the necessary information to proceed with the site design. When designing the site, all the development footprint of the project should be considered as a whole. (A **building footprint** is the area that the building sits on, and the **development footprint** is the sum of all the areas that are affected by the project's activity in the project site. Thus, the development footprint will cover the building footprint and the sidewalks, access roads, hardscapes, parking lots, etc.).

Many people enjoy spending time with pristine views and natural surroundings. In light of this, when designing a development footprint, the project team should always think about preserving open spaces, which will allow the occupants to spend time in the outdoors in a picturesque landscape. In addition, if there are any damaged areas on the project site, including damaged existing vegetation, the project team can work on restoring them to support the environment. Employing these strategies will benefit the environment and society while increasing the project value.

Choosing the type of vegetation to be used in the project is an important decision that will greatly impact outdoor water use consumption. If the project chooses to use **native** or **adapted plants**, they will need less irrigation and maintenance, which will reduce the water use of the project since those plants are accustomed to naturally surviving in that climate.

Native plants (or indigenous plants) are the type of plants that occur and develop naturally in a specific location. Adapted plants are the types of plants that do not occur naturally in a specific location; however, they can nonetheless adapt easily to the climate of the region. In summary, both native and adapted plants can thrive without extensive irrigation, pesticides, and fertilizers.

Other types of plants will require more maintenance and consume more irrigation water, which will result in increased maintenance costs. Plants can be selected to minimize **evapotranspiration**, which is the term used for the return of water to the atmosphere through evaporation from plants. In dry climates, projects can use plants with low evapotranspiration rates, which would require less irrigation water compared with plants with higher evapotranspiration rates.

Invasive plants that exist on-site should be cleared from the site, otherwise they will take over the adjacent existing native and adapted plants. Invasive plants are the types of plants that spread and damage the environment.

Lighting is another important factor that needs to be addressed during the site design. During the green building design, all types of light pollution should be avoided. There are two types of lighting pollution. The first type is the annoying light that intrudes on an otherwise natural or low-light setting, and the second type is the excessive light that leads to discomfort and adverse health effects.

During the lighting design, the project team should eliminate uplighting, glare, overlighting and light trespass (light spilling out of the project boundary) in order to conserve energy and to not create discomfort in the adjacent properties. Direct full cutoff fixtures should point downward to illuminate the project site. The use of shield fixtures will aid in avoiding light trespass and the spilling of light beyond the project boundary.

Innovative approaches can also be discovered for more efficient lighting. An example of this could be the use of **reflective paving materials**, which would help distribute the light across the site; as a result, the number of fixtures needed to illuminate the area would be reduced, as would the building operating costs. For areas that need more lighting, timers can shut off the lighting late at night.

There are other factors to consider when designing a sustainable site, which will be discussed under the Rainwater Management and Heat Island Effect sections.

MINIMIZING CONSTRUCTION IMPACTS

The environmental impacts of construction activities should be minimized, and the necessary strategies should be evaluated before the start of construction.

While grading the project site, the project team should take measures to prevent erosion and sedimentation. To limit erosion and sedimentation, the following methods can be considered:

- Mulching
- Sediment (silt) fencing
- Erosion control blankets
- Berms and constructed ponds
- Seeding
- Straw bales

Sediment fences can be installed around the project site to prevent soil runoff and excessive erosion. The topsoil should always be protected in order to avoid erosion in sloping areas. Projects close to wetlands and water bodies should take extra precautions, and make sure the sensitive areas are not affected by construction activities. Taking constructive action will have a positive effect on local water quality and avoid damage to the infrastructure of the city.

In order to protect the neighborhood, necessary precautions should be taken to prevent dust caused by construction activities from polluting the neighborhood. Those precautions might include cleaning the trucks and other construction vehicles serving the project before leaving the site.

RAINWATER MANAGEMENT

Most people have probably seen the gigantic pipes of storm sewers, which are responsible for carrying stormwater to avoid flooding. But what is the reason that municipalities are working so hard to get rid of this much water?

Think about an undeveloped piece of land with a ground surface of soil. In this scenario, there wouldn't be any flooding. Moreover, rainwater would not be considered as an enemy, but would rather be a best friend.

After we started developing the world, soil was replaced by impervious surfaces, through which the rainwater cannot be absorbed by the soil. Now, all the rainwater that falls on impervious surfaces add up to vast amounts of water that need to be transferred somewhere else. In turn, the transferal of such amounts of rainwater results in other consequences.

With the stormwater runoff, water gets contaminated by harmful chemicals and then flows into natural bodies of water such as seas, oceans, or lakes, resulting in a degraded surface water quality, which consequently harms aquatic life and prevents recreational opportunities.

This type of pollution is also called **nonpoint source pollution** since the point of the pollution source cannot be identified.

Because the rainwater can't be absorbed and filtered by the soil, it cannot replenish the aquifers. To summarize, the natural hydrology of the ecosystem cannot work anymore, and green buildings need to address rainwater management to protect the environment.

The first goal of a green building for rainwater management should be for it to contain the least number of impervious surfaces. Furthermore, the rainwater that falls onto the pervious surfaces should be collected and retained (this is called **on-site water retention**), or at least its flow should be slowed down. If the flow of the rainwater cannot be slowed down, heavy rainfall will overwhelm result in the overwhelming of the municipal storm sewer system.

To minimize impervious surface areas, projects can use vegetated roofs, rain gardens, porous pavements, grid pavers, and additional landscaping in order to allow rainwater absorption through the soil. In parking lots, pervious paving or open-grid paving can be used instead of using asphalt or concrete. For LEED, an impervious surface is a surface that contains less than **50%** perviousness.

Vegetated Roofs *(Photo courtesy of Erik Christensen)*

Rainwater harvesting, another aspect of rainwater management, is a process whereby rainwater is collected and filtered to be reused as an alternative to potable water. The filtered rainwater can be used for irrigation, for toilet and urinal flashing, or even for process water in cooling towers. For storing the filtered rainwater, installing cisterns will be necessary.

Another efficient rainwater management strategy is the use of **low-impact development (LID)**, which is an approach to mimic natural systems and to manage the stormwater closest to its source. LID strategies include decreasing impervious surfaces and increasing vegetation on-site. LID is used to slow down the flow of rainwater out of site and ensures that the rainwater does not get contaminated.

Some strategies of LID include:

- Use of **dry ponds** (detention ponds), which hold the excess rainwater for some time, thereby allowing the rainwater to slowly seep into the ground without contamination. (Dry ponds are excavated areas that detain and slow down stormwater but are dry at other times.)
- Use of **bioswales** and **vegetated filters**, which filter water by having it run through vegetation or through rocks. (Bioswales are a stormwater-control feature that uses a combination of engineered basin, soils, and vegetation.)
- Bioretention

Green infrastructure (**GI**) technologies are accepted in LEED for managing the rainwater as well. Green infrastructure is known to direct the rainwater, collected from impervious surfaces, to vegetation and soil surfaces without routing it to the storm sewer system.

Dry Pond *(Photo courtesy of Michelle Dorothy)*

HEAT ISLAND EFFECT

An example of heat island effect would be the black asphalt surfaces radiating heat on a hot summer day, with the air rising from the asphalt like hot air from a hair dryer. If there was a grass or soil area instead of a black-colored asphalt, as it would have existed before the development, this effect would not occur, as soil cannot absorb the solar heat like black colored asphalt does.

Disturbing the environment with dark-colored, nonreflective surfaces causes the heat island effect. These surfaces absorb heat during hot weather and release it into the atmosphere. Studies show that urban "heat islands" are responsible for 24.2% of global warming.[1] And because of this effect, urban areas can have air temperatures that are 1.8° F to 22° F warmer than the surrounding suburban areas. Higher temperatures will also lead to smog or ground level ozone, which creates consequences for human health.

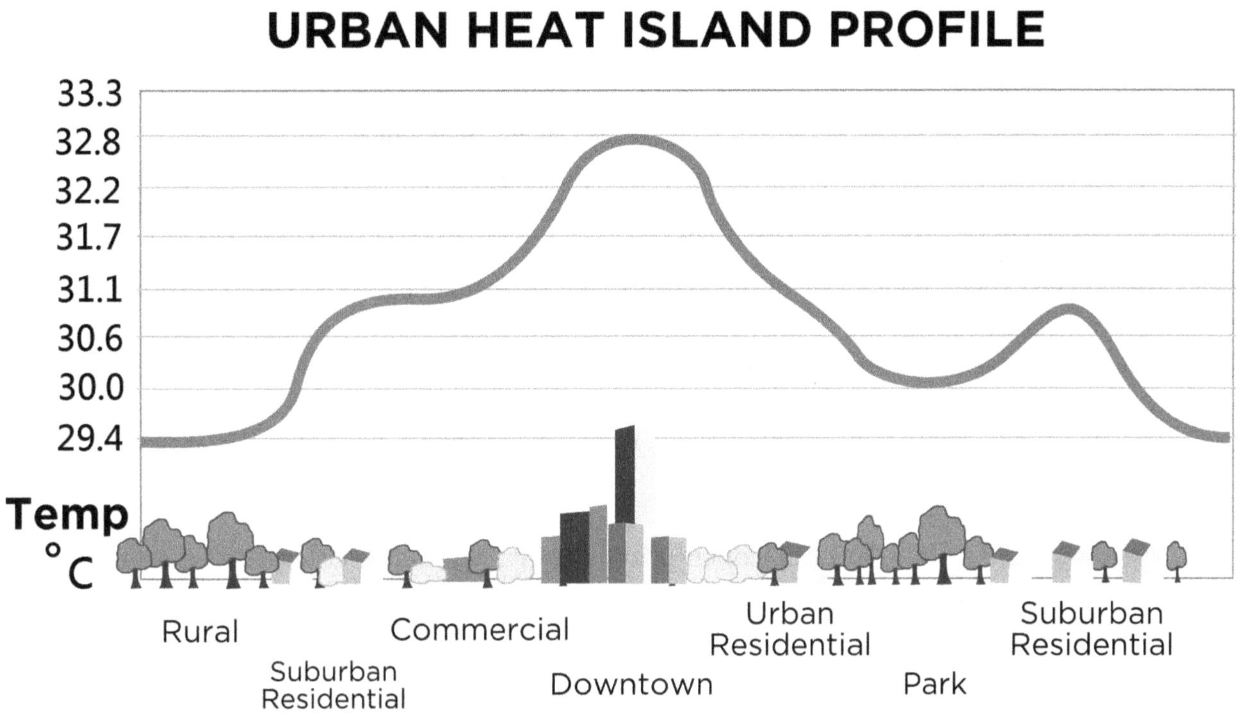

(Photo courtesy of National Oceanic and Atmospheric Administration)

Heat islands are also responsible for increased cooling loads in buildings, which result in higher electricity usage and harm to plants and animals that are sensitive to temperature changes.

Think about wearing a black T-shirt on a hot summer day when walking under the sun at noon, and compare the difference of heat that will be absorbed by the black T-shirt with a white T-shirt.

The black T-shirt will absorb all the heat from the sun while the white T-shirt will reflect back the heat. This can be explained by the differences of their solar reflectance (SR) and solar reflectance index (SRI) values.

A **solar reflectance** (**SR**) value will show the solar energy that is reflected by a surface on a scale of 0 to 1. A black surface will have an SR of 0 while a white surface will have an SR of 1.

A material's **solar reflectance index** (**SRI**) value will indicate a material's ability to stay cool by reflecting solar radiation and emitting thermal radiation. Thus, both the solar reflectance and **emissivity** of a material will be combined to rank the material. Emissivity (infrared or thermal emittance) is a measure that shows how much heat or infrared radiation a material can shed back into the atmosphere. The SRI value of a material is measured from a scale of 0 to 100, and, within that scale, light-colored materials are closer to scoring a 100 SRI while darker-colored materials are closer to scoring 0 SRI. (Lighter-colored materials are sometimes called high **albedo** materials, which is another type of reflectivity measurement.) **Thus, the higher the SRI or SR, the lower the heat island effect.**

In order to classify building materials according to their solar emissions and reflectance, the solar reflectance index (SRI) will be used for the roofing materials, and solar reflectance (SR) values will be used for the nonroof materials (such as hardscape) in LEED credit calculations. In addition to the initial SR and SRI values, aged SR and SRI values will also be needed for LEED credit calculations since the materials' performance will drop as they age.

Following are some strategies to reduce heat island effects:

- Minimize the development footprint, as less development means more exposed soil surface.
- Provide covered or underground parking.
- Use hardscape materials with high solar reflectance (SR) values.
- Reduce the area of paved surfaces exposed to sunlight.
- Install vegetated roofs. Green roofs are more durable than regular roofs and have excellent acoustic and insulation qualities.
- Install **cool roofs**, which reflect more sunlight and absorb less heat than traditional dark-colored roofs.
- Use roof surfaces with high solar reflectance index (SRI) values, such as lighter-colored roofs.
- Provide shade from trees, structures covered by solar panels, or architectural devices with high SR values.
- Use solar panels on the roof.
- Use open-grid paving. In LEED, open-grid pavement should be at least 50% permeable.
- Use **cool pavements**, which reflect more sunlight and absorb less heat than traditional pavements.

Open-Grid Pavement *(Photo courtesy of Immanuel Giel)*

Note that in colder climates, however, using roofing materials with a lower SRI value can actually benefit the project by decreasing heating costs, since the roof will absorb more heat from the sun.

SITE MANAGEMENT

Careful plant selection, innovative irrigation systems, and an efficient outdoor lighting design are some strategies that will aid in sustainable site management. During the site management phase, the following strategies can be implemented:

- Utilize **integrated pest management (IPM)**, which is known as a sustainable approach that combines knowledge about pests, nature, pest prevention, and control methods that minimize pest infestation and damage while minimizing hazards to the building occupants, the property itself, and the environment. Measures should always be taken to first prevent the attraction of pests. Using native and adapted plants for the landscaping can be an effective strategy to combat pests.
- Develop a sustainable site management plan.
- Implement conservation programs with ecologists and nonprofit organizations to protect wildlife species and habitats.
- Use nontoxic chemicals for cleaning outside the building.
- Use nontoxic snow- and ice-removal products.
- Carefully use fertilizers and maintenance products in landscaping.
- Use warmer-colored lights instead of cool blue lights to help reduce light pollution and its associated effects. Cool blue lights may perform well energetically, but they can disrupt natural day-night rhythms for both people and wildlife by resembling morning daylight. This can negatively affect sleep, health, safety, and nighttime visibility, while also contributing to light pollution and reducing visibility of the night sky.

STRATEGIES TO ADDRESS SUSTAINABLE SITES

- Protect and restore habitat.
- Minimize construction impacts.
- Conduct a site assessment to evaluate topography, hydrology, climate, vegetation, soils, human use, and human health effects.
- Preserve open space.
- Use native and adapted plants.
- Minimize hardscape.
- Minimize impervious areas and maximize pervious areas.
- Incorporate stormwater management into the site design.
- Utilize low-impact development (LID) and green infrastructure (GI) strategies to control and redirect stormwater, such as dry ponds, rain gardens, bioswales, and permeable pavement.
- Harvest rainwater.
- Eliminate uplighting, glare, overlighting, and light trespass.
- Use warmer-colored lights instead of cool blue lights to help reduce light pollution and its associated effects.
- Reduce the heat island effect.
- Install reflective roof surfaces.
- Provide covered or underground parking instead of exposed surface parking.
- Reduce the area of paved surfaces exposed to sunlight.
- Plant an urban forest or install a green roof.
- Use products with high SR and SRI values.
- Develop a sustainable site management plan.
- Use nontoxic chemicals for site cleaning.
- Implement integrated pest management (IPM).

SUSTAINABLE SITES PREREQUISITES AND CREDITS

MINIMIZED SITE DISTURBANCE—PREREQUISITE

PREREQUISITE SUMMARY

This prerequisite requires project teams to implement environmental protection measures that minimize construction-related disturbances to the site and neighboring properties, while also protecting and preserving habitat.

To earn the prerequisite, project teams must implement an erosion and sedimentation control (ESC) plan to minimize impacts on soils, rainwater systems, and adjacent properties. In most parts of the world, codes require construction projects to implement an ESC plan to protect the environment and surrounding neighborhoods from pollution generated by construction activities. This prerequisite ensures that, even where no local code exists, disturbances to the site, neighboring properties, and rainwater systems are minimized.

In addition, project teams must conduct a site assessment to understand essential site information and preserve special-status vegetation, protect healthy habitats, and manage invasive vegetation.

PREREQUISITE REQUIREMENTS

Project teams should minimize site disturbance by designing and constructing the project site to meet the following requirements:

1) Erosion and Sedimentation Control Plan

Project teams should develop and implement an erosion and sedimentation control plan for every construction activity tied to the project. The plan must follow the erosion and sedimentation requirements of the **U.S. EPA 2022 Construction General Permit (CGP)**, the **EU Taxonomy: DNSH**, **Pollution Prevention, Item 4 Noise and Dust**; or a local equivalent.

The plan must also include all of the following measures:

- Set up construction-exclusion zones, marked with physical barriers and stormwater controls, to protect any identified critical habitat for threatened or endangered species from discharges and discharge-related activities.
- Perform site inspections for all controls and management practices.
- Take immediate corrective action whenever controls are failing, including repairing or replacing them.

AND

2) Site Assessment

Complete a preconstruction site survey or assessment and use it to guide site design decisions. The assessment should identify special-status vegetation, healthy habitat, and invasive vegetation, and comply with the following requirements:

- Special-status vegetation: Protect 100% of special-status vegetation on-site, as defined by local, state, or federal entities.
- Healthy habitat: Take steps to limit damage to healthy habitat during construction and ongoing operations. Establish exclusion zones with physical barriers to prevent intrusion or disturbance during construction.
- Invasive vegetation: Map existing invasive species on-site and manage their removal and control before and during construction. Specify only native and adapted plants that are not currently listed as invasive.

KEY THINGS TO REMEMBER

- LEED requires all projects to create and implement an erosion and sedimentation control (ESC) plan that follows the erosion and sedimentation requirements of the U.S. EPA 2022 Construction General Permit (CGP), the EU Taxonomy: DNSH, Pollution Prevention, Item 4 Noise and Dust; or a local equivalent.
- LEED requires all projects to protect 100% of special-status vegetation on-site, as defined by local, state, or federal entities.

BIODIVERSE HABITAT—CREDIT

- 1–2 Points (New Construction)

CREDIT SUMMARY

This credit aims to protect on-site greenfield areas during construction and restore damaged habitat. To earn points, project teams must preserve at least 40% of the site's greenfield area from all development and construction activity. To qualify as greenfield, an area must not have been previously developed, graded, or otherwise disturbed.

Project teams can also earn points by restoring previously disturbed land through revegetation. In addition, the credit rewards projects that install bird-friendly glass on the building exterior to help reduce bird collisions.

CREDIT REQUIREMENTS

Option 1: Preserve and Restore Habitat (1–2 Points)

Under this option, project teams have two paths. Under Path 1, the project must preserve 40% of the greenfield area on the site by protecting those areas from all development and construction activity.

Under Path 2, the project must first meet the Path 1 requirement for greenfield protection if greenfield areas exist, and then restore previously disturbed areas of the site if such areas exist. This includes both soil restoration and vegetation restoration. Points are earned based on how much of the previously disturbed area is restored.

For soil restoration, project teams should restore the on-site soils disturbed by previous development and soils disturbed by current construction activities that will serve as a final vegetated area. Imported soils must be reused in a way comparable to their original function, and they may not include the following:

- Soils defined regionally by the **Natural Resources Conservation Service** (**NRCS**) **web soil survey** as prime farmland, unique farmland, or farmland of statewide or local importance.
- Soils from other greenfield sites.
- Sphagnum peat moss or organic amendments that contain sphagnum peat.

For vegetation restoration, project teams should plant native and adapted vegetation that is not currently listed as invasive, including at least 10 species appropriate to the project's **EPA Level III ecoregion** (or local equivalent).

AND/OR

Option 2: Bird-Friendly Glass (1 Point)

To reduce bird collisions, exterior glass installed in the most sensitive zones must have a maximum threat factor of 30, based on the **American Bird Conservancy Threat Factor Database**. This requirement applies to all glass located from grade up to 50 feet (15 meters), up to 20 feet (6 meters) above the finished grade of a green roof, and at any height when the glass is used in guardrails or windshields.

KEY THINGS TO REMEMBER

- Natural Resources Conservation Service designates soil map units as prime farmland, unique farmland, or farmland of statewide or local importance.
- LEED refers to the American Bird Conservancy Threat Factor Database for bird-friendly glass.

ACCESSIBLE OUTDOOR SPACE—CREDIT

- 1 Point (New Construction)

CREDIT SUMMARY

Well-designed outdoor environments can strongly influence the health, comfort, and daily life of building occupants. LEED encourages project teams to create accessible outdoor space that supports everyone, while making it meaningful through elements that promote nature contact, activity, social connection, and community value.

Qualifying projects must dedicate outdoor space equal to at least 30% of the total site area (including the building footprint), and at least 25% of that outdoor area must be vegetated using two or more vegetation types or include an overhead vegetative canopy.

To strengthen daily use and well-being benefits, the credit encourages urban outdoor space features that support nature contact, activity, and social connection. Projects can comply by providing one or more options such as a biophilic space, community garden/urban food production area, recreational area designed for physical activity, or a social area that is pedestrian-focused and includes seating for 5% of building occupants. **Biophilic design** integrates nature into the built environment through direct contact with nature (daylight, plants, water, fresh air, views), natural materials and forms (wood, stone, organic textures and patterns), and nature-inspired spatial layouts. These strategies enhance occupant well-being and indoor environmental quality.

Biophilic design

Finally, the credit supports broader neighborhood value through community-focused outdoor space elements. Projects may include features such as a publicly accessible community space open during daylight hours.

CREDIT REQUIREMENTS

1) Sufficient outdoor space area:

Provide barrier-free, physically accessible outdoor space for people with limited mobility equal to at least 30% of the total site area (including the building footprint). At least 25% of that outdoor space must be vegetated with two or more vegetation types or include an overhead vegetative canopy.

AND

2) Urban outdoor space:

Provide one or more of the following:

- Biophilic space: Outdoor area that satisfies the vegetation restoration requirements of Biodiverse Habitat credit and also includes elements of human interaction such as observation platforms.
- Garden: Area dedicated to community gardens or urban food production.
- Recreational area: Outdoor paving or landscape designed to encourage physical activity, such as courts, fields, a track, play space, or a swimming pool.
- Social area: Pedestrian-focused outdoor space designed for social activities and includes seating for 5% of building occupants.

AND

3) Community outdoor space:

Include one or more of the following elements:

- Community: Space that is open to the public during daylight hours and available to all community members.
- Cultural: Include at least two art installations or sculptures created by local artists.
- Acoustics: Include elements that provide positive soundscapes if the project is within 0.24 miles (400 meters) of a major noise source, such as a roadway or airport.

RAINWATER MANAGEMENT—CREDIT

- 1–3 Points (New Construction)

CREDIT SUMMARY

Before buildings were constructed, ground soils were able to absorb rainwater, and the rainwater was distributed in a natural way. With the development of construction projects, the natural hydrology of ecosystems has been disturbed. Rainwater collected on impervious surfaces is now redirected to other areas, which have begun to receive too much water. At the same time, the soils under impervious surfaces are not receiving any rainwater at all.

This particular credit aims to re-create the rainwater balance by promoting green infrastructure (GI) and low-impact development (LID) rainwater management practices, thus minimizing the disruption caused by the project. To earn the credit, the total volume of rainwater runoff will be calculated based on rainfall data for the project's location, and strategies will be implemented to manage the amount of rainwater.

CREDIT REQUIREMENTS

Option 1: Percentile of Rainfall Events (1–3 Points)

Manage on-site runoff from the developed site for the selected percentile of regional or local rainfall events in a way that best replicates natural site hydrology. The runoff volume for that percentile event must be retained on-site using green infrastructure (GI) and low-impact development (LID) strategies. Retention can be achieved through infiltration, evapotranspiration, or by collecting and reusing stormwater. GI and LID strategies can be either structural or nonstructural. LID and GI practices include rain gardens and bioretention gardens, rainwater harvesting, green roofs, vegetated swales, exfiltration trenches, and permeable pavement.

Points are awarded based on the percentile event retained. Projects retaining runoff from the 80th percentile event earn 1 point, the 85th percentile earns 2 points, and the 90th percentile earns 3 points. For zero-lot-line projects, retaining runoff from the 70th percentile event earns 1 point, the 75th percentile earns 2 points, and the 80th percentile earns 3 points. **Zero-lot-line projects** are those in which the building footprint extends all the way to one or more property (lot) lines, leaving little or no space for setbacks on those sides, except for limited access areas (e.g., entryways, walkways, drive aisles, or service access).

Under this option, project teams can also earn 1 additional point when a portion of the selected percentile event volume is collected and reused to meet one or more end uses for the building and grounds. This can include using the collected rainwater for irrigation, fixture flushing, makeup water systems such as cooling towers or boilers, and other process water demands. Note that collecting and reusing rainwater within the project will also contribute to points earned under the Enhanced Water Efficiency credit.

OR

Option 2: Natural Land Cover Conditions (3 points)

Under this option, project teams should manage the annual increase in on-site runoff volume from the natural land cover condition to the post-developed condition. To do so, project teams first need to identify the site's natural land cover condition and calculate how much rainwater it was able to absorb. Then, project teams should calculate how much rainwater the land can absorb under post-development conditions. The difference between these two values is the additional runoff volume that occurs because of the project's disturbance to the site. Project teams should retain this on-site increase in runoff volume using LID and GI strategies.

For zero-lot-line projects, teams can combine on-site and off-site rainwater strategies to retain runoff from the applicable percentile rainfall event, and can contact local or regional authorities to discuss off-site rainwater strategies, such as participation in community-wide rainwater management programs.

KEY THINGS TO REMEMBER

- LEED addresses a project's rainwater runoff by using on-site green infrastructure and low-impact development strategies, which can include, but are not limited to, rain gardens and bioretention gardens, rainwater harvesting, green roofs, vegetated swales, exfiltration trenches, and permeable pavement.
- To reduce rainwater runoff volume, zero-lot-line projects can contact local or regional authorities to discuss off-site rainwater strategies, such as participation in community-wide rainwater management programs.

ENHANCED RESILIENT SITE DESIGN—CREDIT

- 2 Points (New Construction)

CREDIT SUMMARY

This credit builds on the Climate Resilience Assessment prerequisite and requires project teams to strengthen resilience to at least two of the high-priority hazards identified under that prerequisite by following LEED-defined best practices. If drought is a high-priority hazard for the project's location, the project team can implement drought strategies like meeting enhanced water efficiency targets and using native or drought-tolerant landscaping. For extreme heat, teams can create shaded outdoor areas near entrances, use cooler paving (high solar reflectance or open-grid paving), or add evaporative cooling elements like misters. For flooding, project teams can design critical utilities to be placed above the design flood elevation and use flood-resistant materials below that elevation. For sea level rise, project teams can design elevated foundations, flood barriers, sealants, and coatings to reduce water intrusion.

CREDIT REQUIREMENTS

Project teams should design and construct the site and site structures to meet the following best practices for at least two of the highest-priority hazards identified under Climate Resilience Assessment prerequisite.

1) Drought:

Meet Enhanced Water Efficiency credit, and specify native and/or drought-tolerant adapted plantings so the landscape can survive with less water.

2) Extreme heat:

Include two or more heat-response strategies, such as creating shaded outdoor areas near the building, adding evaporative cooling (misters or fountains), shaping the building to self-shade, providing outdoor cooling stations with backup power, using paving with initial SR ≥ 0.33, and/or using open-grid pavement (at least 50% unbound).

3) Flooding:

Include two or more flood-resilience strategies consistent with ASCE 24 and FEMA 543. Examples include locating critical utilities above the design flood elevation (DFE), floodproofing utilities in retrofits, designing potable water and sewage systems to resist flood damage and backup, anchoring tanks and lines that extend below DFE, and/or using flood-resistant materials below DFE.

4) Hail:

Design and construct the site and site structures to meet **FORTIFIED Commercial High Wind and Hail Specific Design Requirements**, so roofs and exposed elements are better able to withstand hail impacts.

5) Hurricanes and high winds:

In hurricane-prone regions, follow **FORTIFIED Commercial Wind standards.** Strategies may include backup power, storm-ready electrical connections, and windbreaks using landforms, vegetation, and other locally appropriate natural systems.

6) Sea level rise:

Design the site to handle flooding based on sea level rise and storm surge projections over the project's design service life. Add two or more measures such as elevated foundations, water-resistant materials, sealants/coatings to limit water intrusion, flood barriers, GI solutions for storm surge and runoff, backup power to maintain critical functions, integrated drainage systems, and/or retrofits to improve long-term resilience.

7) Tsunamis:

Reduce tsunami impacts through site planning strategies from Designing for Tsunamis (U.S. National Tsunami Hazard Mitigation Program).

8) Wildfires:

Follow wildfire management practices related to wildland–urban interface design, vegetation management, debris disposal, and fire safety for equipment per the **National Wildfire Coordinating Group Standards for Mitigation** (2023).

9) Winter storms:

Include two or more winter-storm strategies, such as ensuring access for vehicles and snow removal equipment, creating a snow removal plan, improving pedestrian safety to parking areas, and selecting native or adapted plantings that can tolerate heavy snow loads.

KEY THINGS TO REMEMBER

- LEED references the FORTIFIED Commercial High Wind and Hail Specific Design Requirements for hail-resilient design.
- LEED references the FORTIFIED Commercial Wind standards for high-wind design.
- LEED references the U.S. National Tsunami Hazard Mitigation Program for tsunami-resilient design.
- LEED references the National Wildfire Coordinating Group Standards for Mitigation to guide wildfire protection measures.

HEAT ISLAND REDUCTION—CREDIT

- 1–2 Points (New Construction)

CREDIT SUMMARY

This credit requires project teams to use strategies that minimize the project's overall contribution to the heat island effect.

Materials that can contribute to the heat island effect include dark-colored roofs, dark-colored pavements, asphalt surfaces, and other low-reflectance hardscape. This credit awards points for minimizing a project's heat island effect by using vegetated roofs, high-reflectance roofs, light-colored nonroof measures, and/or providing covered parking. For the material reflectance calculations, solar reflectance index (SRI) values are used for roofs, and solar reflectance (SR) values are used for nonroof measures.

The most effective way to reduce heat island impacts is to minimize hardscape and increase tree canopy as much as possible. Shaded surfaces do not absorb and re-radiate heat the same way bare asphalt or concrete does, and trees cool the air through shade and evapotranspiration. Neighborhood tree cover also matters: areas with more trees are typically cooler than those with little or no tree cover, which reduces the overall heat island effect.

According to **American Forests**, a map of tree cover in many U.S. cities is too often also a map of race and income. With few exceptions, trees are sparse in socioeconomically disadvantaged neighborhoods and more prominent in wealthier ones. For that reason, they created the Tree Equity Score. **Tree Equity Score** highlights the inequities in tree cover and identifies where trees are needed most in each community, helping address disparities by focusing resources in neighborhoods with the highest need. In Option 3 of this credit, LEED also addresses this inequity by awarding a point to project teams that check the project site's Tree Equity Score and increase on-site tree canopy cover if the area is ranked high priority or highest priority.

For places without a Tree Equity Score, tools such as Global Forest Watch can help teams understand neighborhood tree cover.

CREDIT REQUIREMENTS

Option 1: Nonroof and Roof (1 Point)

Meet the following criteria:

$$\frac{\text{Area of Non-roof Measures}}{0.5} + \frac{\text{Area of High-Reflectance Roof}}{0.75} + \frac{\text{Area of Vegetated Roof}}{0.75} \geq \text{Total Site Paving Area} + \text{Total Roof Area}$$

Project teams are free to use any combination of nonroof, high-reflectance roof, and vegetative roof strategies to satisfy the credit's criteria, provided that the weighted sum of site design strategies is greater than or equal to the sum of the site's total pavement and roof areas. The details of each measure are as follows.

Nonroof measures:

- Shade pavement areas with new or existing plants, measured in plan view at noon, assuming a 10-year canopy width, or with vegetated structures. Plants or vegetated structures must be in place at the time of the occupancy permit.
- Structures covered by energy-generation systems (e.g., solar thermal collectors, photovoltaics, wind turbines, etc.).
- Architectural devices or structures. If the architectural device or structure is a roof, it should have an aged SR of at least 0.28, measured in accordance with ANSI/CRRC S100. If the architectural device or structure is not a roof, or if aged SR information is not available, it should have an initial SR of at least 0.33, measured in accordance with ANSI/CRRC S100.
- Paving materials with an initial SR of at least 0.33.
- An open-grid pavement system that is at least 50% unbound.

High-reflectance roof:

Use roofing materials with an aged SRI value equal to or greater than the values in the following table. If an aged SRI value is not available, the roofing material should have an initial SRI equal to or greater than the values in the following table.

	Minimum solar reflectance index value by roof slope		
	Slope	**Initial SRI**	**Aged SRI**
Low-sloped roof	≤ 2:12	82	64
Steep-sloped roof	> 2:12	39	32

Vegetated roof:

Install a vegetated roof with native or adapted plant species.

AND/OR

Option 2: Parking Under Cover (1 Point)

Place all parking spaces under cover. Any roof used to shade or cover parking must meet one of the following:

- Have an aged SRI of at least 32. If an aged SRI value is not available, the material used should have an initial SRI of at least 39 at installation.
- Be a vegetated roof.
- Be covered by energy-generation systems, such as solar thermal collectors, photovoltaics, or wind turbines.

AND/OR

Option 3: Tree Equity (1 Point)

For U.S. projects, project teams should check the American Forests Tree Equity Score for the project site's location. If the area is ranked high priority or highest priority, the results should be used to guide an increase in on-site tree canopy cover compared with the existing condition.

For international projects, teams should refer to the Human Impact Assessment prerequisite and evaluate tree cover both on-site and in the surrounding community, either by using a local tree census or by conducting a site assessment. The results of this evaluation should then be used to inform an increase in on-site canopy cover accordingly. International projects may use platforms such as **Global Forest Watch** to understand tree coverage in the project neighborhood. Note that projects with no exterior work are exempt from these requirements.

KEY THINGS TO REMEMBER

- For heat island effect calculations, LEED uses the initial SR and SRI, as well as the aged SR and SRI values.
- For nonroof measures, SR values are used; for roofs, SRI values are used.
- The use of lighter-colored surfaces reduces the heat island effect, and higher SR and SRI values mean a lower heat island effect.
- Note how using energy generation systems such as solar photovoltaics can contribute to several LEED credits, including the Heat Island Reduction credit, by providing shade.
- LEED references the American Forests Tree Equity Score to prioritize adding on-site tree canopy in projects located in high-priority and highest-priority ranked areas. International projects can use Global Forest Watch to assess neighborhood tree cover.

LIGHT POLLUTION REDUCTION—CREDIT

- 1 Point (New Construction)

CREDIT SUMMARY

Light pollution is another factor that creates environmental problems. Overexposure to artificial lighting causes sleeping problems in humans and disorientation for birds, and light pollution is a waste of energy that can bring further environmental impacts. Minimizing light pollution improves human health and protects wildlife.

This credit mandates efficient lighting design by setting requirements for uplight and light trespass (backlight and glare) for exterior luminaires inside the project boundary as well as requirements for internally illuminated exterior signage. **Uplight** is light emitted above the horizontal plane that contributes to sky glow; **backlight** is the light emitted behind a luminaire and is a primary contributor to light trespass onto adjacent sites; and **glare** is excessive brightness—often from high-angle light—that can cause visual discomfort and reduce visibility.

The **BUG rating method** is also referenced in the credit which is a luminaire classification system that classifies a luminaire according to backlight, uplight, and glare.

CREDIT REQUIREMENTS

Project teams should meet the following uplight, light trespass, and internally illuminated exterior signage requirements for all exterior luminaires located inside the project boundary, based on the photometric characteristics and the lighting zone of the project. The photometric characteristics of each luminaire should reflect its orientation and tilt.

To determine the lighting zone of the project, project teams should refer to the **Illuminating Engineering Society and Dark Sky International (IES/IDA) Model Lighting Ordinance (MLO) User Guide**. Based on this guide, there are five lighting zones—LZ0, LZ1, LZ2, LZ3, and LZ4—with appropriate lighting levels for each. LZ0 is for areas where the natural environment would be negatively affected by lighting, while LZ4 is for areas where high light levels are needed, like Times Square in New York City. In this credit, project teams will first choose the lighting zone that is appropriate for their project's location and then work on fulfilling the credit requirements.

The lighting that is exempt from these requirements includes:

- Directional lighting for transportation
- Lighting for theatrical purposes (stages, etc.)
- Hospital emergency department and helipad lighting
- National flag lighting

Uplight:

All exterior luminaires within the project boundary should meet the LEED's defined uplight ratings based on the lighting zone. An uplight rating of U0 indicates that the fixture emits zero light upward into the night sky and prevents it from contributing to sky glow.

Using fully shielded fixtures that direct light downward will reduce excess light pollution.

AND

Light Trespass:

As its name implies, light trespass occurs when unwanted artificial light spills into areas where it is not intended, such as street lights illuminating adjacent properties or shining into other properties' windows.

To meet the credit's light trespass requirements, project teams should not exceed the luminaire backlight and glare ratings, as defined in **IES TM-15-11 Addendum A** (BUG ratings), based on the mounting location and distance from the lighting boundary. The lighting boundary is generally located at the project property lines. However, certain conditions can modify the lighting boundary as follows:

- If the property line is adjacent to a public area that is a parking lot, walkway, bikeway, or plaza, the lighting boundary can be moved 5 feet (1.5 meters) beyond the property line.
- If the property line is adjacent to a street, alley, or transit corridor, the lighting boundary can be moved to the centerline of the street, alley, or transit corridor.
- If the adjacent properties are owned by the same entity.

AND

Internally Illuminated Exterior Signage:

Do not exceed LEED's maximum luminance level for internally illuminated signage during nighttime hours.

KEY THINGS TO REMEMBER

- Know the definitions of uplight, backlight, and glare.
- To address uplight, project teams should use fully shielded fixtures that direct light downward, which also helps prevent light trespass. To address light trespass, project teams should avoid poorly shielded lighting located too close to property lines.
- To determine the project's lighting zone, LEED refers to the Illuminating Engineering Society and Dark Sky International (IES/IDA) Model Lighting Ordinance (MLO) User Guide.
- For light trespass, LEED refers to IES TM-15-11 Addendum A (BUG ratings), based on the mounting location and distance from the lighting boundary.
- Directional lighting for transportation, lighting for theatrical purposes (stages, etc.), hospital emergency department and helipad lighting, and national flag lighting are exempt from LEED's lighting requirements.

END OF CHAPTER EXERCISES

EXERCISE: MATCHING GAME

Match the following terms to their definition.

Terms		
Native Plants	Low-Impact Development (LID)	Solar Reflectance (SR)
Invasive Plants	Bioswales	Solar Reflectance Index (SRI)
Nonpoint Source Pollution	Integrated Pest Management (IPM)	

1) ________________________: Shows the solar energy that is reflected by a surface on a scale of 0 to 1.

2) ________________________: Indicates a material's ability to stay cool by reflecting solar radiation and emitting thermal radiation. Thus, both the solar reflectance and emissivity of a material will be combined to rank the material.

3) ________________________: A sustainable approach that combines knowledge about pests, nature, pest prevention, and control methods that minimize pest infestation and damage while minimizing hazards to the building occupants, the property itself, and the environment.

4) ________________________: Plants that occur and develop naturally in a specific location.

5) ________________________: Plants that spread and damage the environment by taking over the adjacent existing native and adapted plants.

6) ________________________: Type of pollution in which its source cannot be identified and which generally results from multiple sources.

7) ________________________: A rainwater management strategy that mimics natural site hydrology by managing stormwater close to where it falls.

8) ________________________: A stormwater-control feature that uses a combination of engineered basin, soils, and vegetation.

ANSWER: MATCHING GAME

1) Solar Reflectance (SR)	4) Native Plants	7) Low-Impact Development (LID)
2) Solar Reflectance Index (SRI)	5) Invasive Plants	8) Bioswales
3) Integrated Pest Management (IPM)	6) Nonpoint Source Pollution	

PRACTICE EXAM

To access the end-of-chapter practice exam online in our exam simulator, scan the QR code below with your laptop, phone, tablet, or another compatible device, then log in using your email address and password. If you have trouble scanning the QR code, visit www.projectific.com/account and log in using your email address and password. If you have not yet created a password, please refer to the *How to Access the Exam Simulator* section at the back of the book for instructions.

Chapter 7: Water Efficiency

Seventy percent of the earth's surface is covered by water; however, fresh water only comprises 1% of the total volume and is the only percentage that is accessible for human use. According to the United Nations, a quarter of the world's children could be living in places with extreme water shortages by 2040.

With the increase of residential, commercial, and industrial development, the use of potable water increases. From flushing toilets to industrial water uses, potable water is becoming wastewater, which overwhelms treatment facilities and results in heavy energy use during water treatment, as well as creation of more greenhouse gases.

Green buildings need to address this issue by improving their water efficiency. Note that "efficiency" does not mean finding an alternative method; rather, it means doing the same thing by using less of the same resource. As a result, water will still be used, but it will be used in a smarter way.

The **water balance** approach, which aims to balance water supply with water consumption, is not achievable for every location. (An example of the water balance approach would be to use only captured rainwater and/or underground water for the water needs of the building, so no water would be used from municipal lines.) Nonetheless, getting as close as possible can still bring about a considerable change.

This chapter will discuss both the importance of water efficiency and its effect on the green building design. After discussing the WE credit category, we will go over all the WE prerequisites and credits individually to show how these features are used in LEED certification for a LEED BD+C: New Construction and Major Renovations project.

The WE credit category will be discussed under two major sections:

- Indoor water use
- Outdoor water use

However, before discussing the finer details, we will mention the required LEED calculations for both indoor and outdoor water use, as these are used in the WE prerequisites and credits.

LEED INDOOR AND OUTDOOR WATER USE CALCULATIONS

1) Indoor water use calculations:

When approaching the strategies to reduce the water consumption of a project, it is necessary to first calculate a **baseline** water usage and then find innovative ways to establish water use reductions from the baseline during the design phase. In LEED, projects will be awarded points according to the percent reduction made from their baseline water usage.

To calculate the baseline water usage of a project, first the baseline rates for flow and flush fixtures and fittings should be determined. The occupant usage should then be calculated to find out how much those fixtures and fittings will be used. This calculation is made to demonstrate that the more occupants there and the more time they spent in a building, translate to greater total water usage.

Finally, project teams will need to discover innovative strategies for reducing water consumption as much as possible.

2) Outdoor water use calculations:

The projects that contain landscaping will need to calculate the baseline **landscape water requirement (LWR)**. LWR is the amount of water that the landscape will require during the site's peak watering month.

To calculate the LWR, projects will use an USGBC-approved calculator which the project team needs to input the types of vegetation, the area of each vegetation, project location, and irrigation system type. With this information, the calculator will determine the total LWR of the project which will be the baseline value. Next, project teams will need to discover innovative ways for reducing this value.

Projects without a landscape will be exempt from outdoor water use calculations.

INDOOR WATER USE

Use of water for urinals, toilets, showers, kitchen, and other applications all contribute to indoor water use. Indoor potable water consumption can be reduced by using water-efficient fixtures and fittings and/or using nonpotable water sources where appropriate.

The use of **WaterSense**-labeled products can be a good strategy that will contribute to reducing indoor water usage. WaterSense is a program developed by the U.S. Environmental Protection Agency to identify high-performance, water-efficient fixtures and fittings. A fixture or a fitting that has a WaterSense label will guarantee high water efficiency.

Following are different types of efficient fixtures and fittings that can be installed to decrease potable water consumption:

- Dual-flush toilets: This type of toilets have two buttons, one for full flush, the other one for half flush, depending on the type of waste.
- High-efficiency toilets (HETs): WaterSense-labeled toilets, which save 20% of the water typically used in every flush.
- Waterless urinals: These types of urinals do not need any water and have lower maintenance costs.
- Composting toilet systems: Such systems do not require any water since they treat waste with a microbiological process.
- Low-flow showerheads and faucets
- Faucets with low-flow aerators and/or motion sensors

In addition to high-efficiency fixtures, the use of harvested rainwater and/or graywater as the flush water of toilets can decrease potable water consumption.

Graywater is the untreated household water that does not come into contact with toilet waste. Used water from bathtubs, showers, bathroom washbasins, and clothes washers and laundry tubs can be examples of graywater and may be used as flush water in toilets or urinals.

By installing graywater treatment systems, the used water can be filtered and then sent back to the building for reuse. The definition of graywater changes from state to state, and some states do not allow the use of graywater as toilet flush water. Because of this, it is advised that the project team first check with the local code requirements before deciding on using graywater.

It is also important to mention blackwater and reclaimed water for the purposes of the exam. **Blackwater** is the term to describe the used water that has come into contact with waste. Thus, the water collected from the urinals and toilets can be classified as blackwater. **Reclaimed water** is former blackwater that has been treated and purified for reuse and can even be used for irrigation. However, projects again need to be verified according to local code requirements, as some codes may classify some types of reclaimed water as blackwater and may not allow its use.

Reducing **process water** usage is also essential in reducing potable water usage. Process water is the type of water used by mechanical or other types of systems in buildings such as cooling towers or medical equipment in hospitals. Using reclaimed water, harvested rainwater, or graywater as process water can make a great contribution to reducing potable water consumption. If the projects also choose to install **closed-loop** systems, which use and circulate the same process water inside, instead of using **open-loop** systems, then even more potable water can be saved. Open-loop systems require outside water (which is called the **makeup water**) to support the system.

Installing submeters would be another plus for the projects to track and log water use trends and check fixture performance. Further, it would allow for the discovery of innovative ways to reduce potable water usage. Without such data, it would not be possible to determine the potable water use of individual building systems, pinpoint items that consume large amounts of potable water, or detect leaks.

When doing water-use calculations for LEED, the water usage of **flush fixtures** such as toilets and urinals is measured in **gallons per flush** (**gpf**). For **flow fixtures**, such as sink faucets, showerheads, and aerators, the water usage is measured in **gallons per minute** (**gpm**).

OUTDOOR WATER USE

In some projects, potable water used for irrigation exceeds the total indoor water consumption of the building. Water reductions in irrigation can be established through water-wise landscaping designs, water-efficient irrigation technologies, nonpotable water use (such as using reclaimed water, graywater, or harvested rainwater), and installation of submeters to track and log irrigation trends.

As mentioned before, careful plant selection will make a great contribution to reducing outdoor water use. The use of native and adapted plants, or **xeriscaping** (type of landscaping that does not need any irrigation), and clearing out any invasive plants can help considerably. Areas covered with turf or grass will need high amounts of irrigation water. Project teams should evaluate the related LEED prerequisites and credits before deciding on their use. **Mulching** around a plant, which is a protective layer applied to the surface of soil, will help to keep the roots of the plants cool and therefore prevent evaporation.

Using high-performance irrigation technologies such as **drip irrigation systems**, **bubbler distribution systems**, and **channel water directly to root systems** are other strategies to consider. The use of **weather-based irrigation controllers**, to respond to weather conditions, will avoid watering the vegetation while it is raining.

Drip irrigation systems are the types of microirrigation systems that drip water to the roots of plants to minimize the use of irrigation water and fertilizers. They are the <u>most water-efficient systems</u> and have very short payback periods. Scheduling of the irrigation is important, as it is best to water the landscape in the early morning, which is the coolest part of the day.

For the projects that install a landscape that does not require any permanent irrigation, the period after the first two years of the new planting will be considered in the LEED prerequisites and credits, as the newly planted landscape will need more water to become established in the first two years.

Xeriscaping *(Photo courtesy of Jeffrey Beall)*

STRATEGIES TO ADDRESS WATER EFFICIENCY

- Install efficient plumbing fixtures such as dual-flush toilets, high-efficiency toilets (HET), waterless urinals, composting toilet systems, low-flow showerheads and faucets, and faucets with low-flow aerators and/or motion sensors.
- Use nonpotable water such as reclaimed water, harvested rainwater, and graywater as indoor and outdoor water.
- Install submeters to individually track water consumption and detect leaks.
- Use native or adapted plants and avoid invasive plants.
- Use xeriscaping.
- Install high-efficiency irrigation systems such as drip irrigation systems, bubbler distribution systems, and systems that channel water directly to root systems with weather-based irrigation controllers.

WATER EFFICIENCY PREREQUISITES AND CREDITS

WATER METERING AND REPORTING—PREREQUISITE

PREREQUISITE SUMMARY

What gets measured gets managed. Without tracking building water use, there is no way to tell whether the building is performing as designed or quietly wasting water through leaks, inefficient fixtures, or poor operations.

This prerequisite focuses on measuring total water consumption so project teams can better manage potable water resources, strengthen day-to-day water management, and spot new opportunities to save water over time. To comply, project teams must install permanent water meters to monitor, record, and report whole-project water use for both potable water and alternative water sources serving the building and associated grounds.

The meter data must be accessible to the facility manager and/or tenants, and alternative water must be metered separately from municipally supplied potable water.

PREREQUISITE REQUIREMENTS

Project teams should install or use existing permanent water meters to monitor, record, and report the total water consumption for potable and alternative water sources for the building and associated grounds.

Report whole-project water use for each type of water source serving the building and its associated grounds, and confirm the following:

- The facility manager and/or tenant(s) can access the meter data.
- Alternative water sources are metered separately from municipally supplied potable water.
- The project commits to sharing whole-project water use data with USGBC at least once per year. This commitment must continue for five years, or until the building changes ownership or lessee.

KEY THINGS TO REMEMBER

- USGBC requires all projects to install or use existing permanent water meters to monitor, record, and report the total water consumption for potable and alternative water sources for the building and associated grounds.
- USGBC requires all projects to share whole-project water use data with USGBC at least once per year. This commitment must continue for five years, or until the building changes ownership or lessee.

MINIMUM WATER EFFICIENCY—PREREQUISITE

PREREQUISITE SUMMARY

Water is our main resource for living, and it should not be used at a faster rate than it can be replenished. Unfortunately, in many cities, water levels are rapidly decreasing, which is likely to cause many serious problems in the near future. Moreover, every gallon of drinking water delivered to a building goes through treatment, pumping, and distribution—using energy and creating carbon emissions.

To comply with this prerequisite, projects must meet minimum efficiency requirements for fixtures and fittings by complying with the maximum specified flush and flow rates. Alternatively, project teams can pursue Option 2, which requires at least a 20% reduction in aggregate fixture and fitting water use compared with the baseline water use. Projects in the United States can use **WaterSense** labeled products for toilets, urinals, private lavatory faucets, and showerheads, as WaterSense products are verified to be water-efficient.

Next, the prerequisite addresses equipment water efficiency, as equipment can consume more water than all the other water-consuming items in the building. Any newly installed appliances and equipment within the project boundary must meet LEED's minimum water efficiency requirements. For residential and commercial clothes washers, residential dishwashers, and ice machines, project teams should use **ENERGY STAR** labeled equipment or equivalent performance.

Finally, the prerequisite addresses outdoor water use. In nonagricultural water use, landscape irrigation can account for 30% to 70% of water consumed. As a result, this prerequisite promotes the smart use of outdoor water by choosing landscaping that requires no irrigation (xeriscaping), or project teams will need to reduce irrigation water needs by at least 30% from the calculated baseline through plant selection and irrigation system efficiency.

PREREQUISITE REQUIREMENTS

Meet the minimum efficiency requirements for fixtures and fittings, equipment, and outdoor use as outlined below.

1) Minimum Fixture and Fittings Efficiency

Meet the minimum efficiency requirements for fixtures and fittings by choosing one of the following options.

Option 1: Prescriptive Path—Maximum Flush and Flow Rates
For all new and existing fixtures and fittings within the project boundary, projects should not exceed the maximum flush and flow rates listed in the following table.

Fixture or Fitting	Maximum Installed Flush or Flow Rate (IP units)
Toilet (water closet)	1.28 gpf
Urinal	0.5 gpf
Public lavatory faucet	0.5 gpm
Private lavatory faucet	1.5 gpm
Kitchen faucet	1.8 gpm
Showerhead	2 gpm

Projects located in the United States should use WaterSense-labeled products for toilets, urinals, private lavatory faucets, and showerheads.

OR

Option 2: Performance Path—Calculated Reduction
For all new and existing fixtures and fittings within the project boundary, project teams should reduce the project's aggregate water use by at least 20% from the baseline listed in the following table.

Fixture or Fitting	Maximum Installed Flush or Flow Rate (IP units)
Toilet (water closet)	1.6 gpf
Urinal	1 gpf
Public lavatory faucet	0.5 gpm
Private lavatory faucet	2.2 gpm
Kitchen faucet	2.2 gpm
Showerhead	2.5 gpm

2) Minimum Equipment Water Efficiency

Any newly installed appliances, equipment, and process loads within the project boundary must meet LEED's minimum water efficiency requirements. For residential and commercial clothes washers, residential dishwashers, and ice machines, project teams should use ENERGY STAR–labeled equipment or equivalent performance.

3) Minimum Outdoor Water Use Efficiency

Meet the minimum outdoor water use efficiency requirements by choosing one of the following options.

Option 1: No Irrigation
Show that the landscape does not need a permanent irrigation system beyond a maximum two-year establishment period. Projects with no landscape area achieve this option without further action.

OR

Option 2: Efficient Irrigation
Reduce irrigation water needs by at least 30% from the calculated baseline through plant selection and irrigation system efficiency.

KEY THINGS TO REMEMBER

- To address indoor water use, LEED requires all projects to meet minimum efficiency requirements for fixtures and fittings by complying with the maximum specified flush and flow rates or by reducing the project's aggregate water use by at least 20% through the use of more efficient fixtures and fittings.
- Projects in the United States can use WaterSense-labeled products for toilets, urinals, private lavatory faucets, and showerheads, as WaterSense products are verified to be water-efficient.
- For residential and commercial clothes washers, residential dishwashers, and ice machines, LEED requires project teams to use ENERGY STAR–labeled equipment or equivalent performance.
- LEED permits temporary irrigation during a two-year establishment period for landscaping without irrigation needs.
- All projects with irrigation needs should reduce baseline outdoor water use by 30% from the calculated baseline through plant selection and irrigation system efficiency.

WATER METERING AND LEAK DETECTION—CREDIT

- 1 Point (New Construction)

CREDIT SUMMARY

This is the credit of the Water Metering and Reporting prerequisite, which awards a point for projects that submeter the water use of different subsystems or install a leak detection system that can identify leaks and alarm upon leak detection.

By submetering different building systems, a facility manager can identify major water uses and look for ways to reduce water use. With a leak detection system, on the other hand, the facility manager and/or tenant(s) can spot leaks early and avoid waste and costly damage later on.

The credit also requires the submetering data and leak detection system data to be accessible in real time through a local network, building management system, cloud service, app, or online database.

CREDIT REQUIREMENTS

Option 1: Submeters (1 point)

Install permanent water meters for each of the following applicable subsystems:

- Indoor plumbing fixtures and fittings: Meter systems serving at least 80% of indoor fixtures and fittings as described in the Minimum Water Efficiency prerequisite.
- Irrigation system
- Each makeup water system, such as for cold water inlet for domestic hot water, swimming pools, chilled water systems, and process water systems.
- Commercial kitchen
- Laundry

The facility manager and/or tenants should be able to access submeter data in real time via a local network, building management system, cloud service, app, or online database. Additionally, all submeters must be capable of recording data at least hourly.

For residential projects, teams should install a permanent water meter for each dwelling unit that measures total potable water use for that unit.

OR

Option 2: Leak Detection Sensors (1 point)

Install permanent water flow meters for each of the following applicable subsystems:

- Project irrigation system at the point of entry.
- At least 50% of flush fixtures.
- Each makeup water system, such as cold water inlet for domestic hot water, swimming pools, chilled water systems, and process water systems.

The leak detection system must be able to detect leaks either by identifying abnormal flow above the normal range or by physically detecting water leakage, and it must initiate an alarm when a leak is detected. The facility manager and/or tenant(s) must be able to access sensor data in real time via a local network, building management system, cloud service, app, or online database.

Project teams should also develop an action plan that explains how the building manager or tenant will access data in real time and how leaks will be addressed and repaired.

KEY THINGS TO REMEMBER

- With this credit, LEED awards projects that submeter the water use of different subsystems or install a leak detection system that can identify leaks and alarm upon leak detection.

ENHANCED WATER EFFICIENCY—CREDIT

- 1–8 Points (New Construction)

CREDIT SUMMARY

This credit builds on the Minimum Water Efficiency prerequisite and encourages projects to reduce potable water use even further. The prerequisite sets the "minimum standard," but this credit is where project teams can make significant reductions through smarter design choices by reducing indoor and outdoor demand, improving process water performance, and, when possible, replacing potable water with alternative water sources. These sources can include, but are not limited to, municipally supplied reclaimed water, graywater, rainwater, stormwater, treated seawater, condensate, foundation dewatering water, used process water, and reverse osmosis reject water.

CREDIT REQUIREMENTS

This credit offers up to 8 points, and project teams may either pursue Option 1 or use any combination of Options 2–6 to reach the point total.

Option 1: Whole-Project Water Use (1–8 points)

Develop a baseline water use and a proposed water use model to earn points based on the overall percent reduction from baseline. A greater percent reduction will grant more points.

OR

Option 2: Fixture and Fittings—Calculated Reduction (1–3 points)

Further reduce indoor fixture and fitting water use beyond the calculated baseline established in the Minimum Water Efficiency prerequisite, Path 2: Performance Path—Calculated Reduction. Points are based on the percentage reduction. Additional potable water savings beyond the prerequisite can also be achieved using alternative water sources.

AND/OR

Option 3: Appliance and Process Water (1–2 points)

New appliances and process equipment installed within the project boundary must meet LEED's efficiency requirements for their equipment type.

AND/OR

Option 4: Outdoor Water Use (1–2 points)

Under this option, project teams choose one of two paths. If the landscape does not require a permanent irrigation system beyond a maximum two-year establishment period, project teams automatically satisfy Path 1. In Path 2, on the other hand, project teams can reduce irrigation water needs by 50% or 100% from the calculated baseline through plant selection and irrigation system efficiency.

AND/OR

Option 5: Optimize Process Water Use (1–2 points)

This option targets process water, especially cooling towers. Project teams can earn points by limiting cooling tower cycles, reducing cooling water use, or using alternative water for process uses, with points based on how much alternative water is used.

AND/OR

Option 6: Water Reuse (1–2 points)

Under this option, project teams choose one of two paths. In Path 1, project teams install a water supply system that is reuse-ready, meaning reclaimed or alternative water can be delivered to one or more eligible end uses, with space provided for treatment equipment if needed. In Path 2, project teams actually use reclaimed water by providing either an on-site water reuse system or municipally supplied reclaimed water for indoor, outdoor, and/or process water demands. Eligible end uses include irrigation, flushing fixtures, makeup water systems (such as cooling towers or boilers), and other process water uses.

KEY THINGS TO REMEMBER

- To reduce potable water use, LEED encourages projects to use alternative water sources, including but not limited to municipally supplied reclaimed water, graywater, rainwater, stormwater, treated seawater, condensate, foundation dewatering water, used process water, and reverse osmosis reject water.
- Cooling towers use a significant amount of potable water. To reduce potable water use, projects that include cooling towers should consider opportunities to reduce concentration cycles, optimize water use for cooling, or use alternative water sources.

END OF CHAPTER EXERCISES

EXERCISE: MATCHING GAME

Match the following terms to their definition.

Terms		
Landscape Water Requirement (LWR)	Reclaimed Water	Closed-Loop Systems
Graywater	Process Water	
Blackwater	Xeriscaping	

1) ____________________: Former blackwater that has been treated and purified for reuse.

2) ____________________: Water used by mechanical or other types of systems in buildings such as cooling towers or medical equipment in hospitals.

3) ____________________: Type of landscaping that does not need any irrigation.

4) ____________________: The amount of water that the landscape of the site will require during the site's peak watering month.

5) ____________________: Untreated household water that did not come into contact with toilet waste.

6) ____________________: Used water that has come into contact with waste.

7) ____________________: Systems that circulate the same process water inside.

ANSWER: MATCHING GAME

1) Reclaimed Water	4) Landscape Water Requirement (LWR)	7) Closed-Loop Systems
2) Process Water	5) Graywater	
3) Xeriscaping	6) Blackwater	

PRACTICE EXAM

To access the end-of-chapter practice exam online in our exam simulator, scan the QR code below with your laptop, phone, tablet, or another compatible device, then log in using your email address and password. If you have trouble scanning the QR code, visit www.projectific.com/account and log in using your email address and password. If you have not yet created a password, please refer to the *How to Access the Exam Simulator* section at the back of the book for instructions.

CHAPTER 8:

ENERGY AND ATMOSPHERE

This credit category contains the **highest total points** among the LEED credit categories (**33 points**) because energy use and the burning of fossil fuels are the biggest contributors to global warming. Green buildings need to use less energy and rely on sustainable energy sources rather than fossil fuels. When fossil fuels are burned to produce electricity, they generate harmful greenhouse gas (GHG) emissions, in other words, carbon emissions. As discussed in the introductory chapters, greenhouse gases trap solar energy (heat) in Earth's atmosphere and prevent that heat from escaping back into space. This is called the **greenhouse effect**.

Even though there are numerous greenhouse gases in Earth's atmosphere, such as nitrous oxide, methane, and ozone, carbon dioxide is the main GHG contributing to present-day climate change. As a result, we began expressing the total emissions of all GHGs using a metric called **CO_2 equivalent**.

Extracting and burning fossil fuels harms water, land, and air while also intensifying climate change. In water systems, fossil fuels and climate change contribute to ocean warming, acidification, ecosystem destruction, melting ice caps, and stronger storms. On land, climate change is disrupting ecosystems, increasing insect infestations, threatening crop productivity, and worsening flooding and soil erosion. In the air, burning fossil fuels releases both greenhouse gases and harmful pollutants that damage outdoor and indoor air quality, increasing risks to both environmental and human health.

Unfortunately, electrical grids still get **60% of their energy from fossil fuels**. On average, more than half of the energy in the original fossil fuels is lost as heat when it is converted into electricity. On top of that, a significant portion is also lost when electricity is transmitted over long distances from power plants to buildings.

According to the International Energy Agency, the energy used in buildings accounts for **30% of total energy consumption** and **26% of energy-related carbon emissions worldwide**. As a result, reducing building energy use through smart strategies and electrifying building systems to avoid all on-site combustion of fuels can make an enormous contribution to decarbonization. Electrification refers to replacing non-electric energy sources with electricity.

Similar to the water-efficiency credit category, in which projects calculate baseline water usage and establish reductions by employing water-efficient strategies, this credit category requires calculating the baseline energy usage of the whole building, allowing teams to identify innovative ways to reduce building energy use.

The New Buildings Institute study also found that a significant percentage of buildings underperformed relative to their benchmarks. Commissioning building systems and monitoring their performance are therefore very important to ensure that established benchmarks translate into actual performance, which is also addressed in this credit category.

This chapter will discuss both the importance of the Energy and Atmosphere credit category and its effect on green building design. After discussing the EA credit category, the book will go over all the EA prerequisites and credits individually to show how these features are used in LEED certification for a LEED BD+C: New Construction and Major Renovations project.

The EA credit category will be discussed under five major sections:

- Energy demand
- Use of refrigerants
- Energy efficiency
- Electrification
- Renewable energy
- Power resilience
- Ongoing performance

ENERGY DEMAND

Green building projects need to set goals for saving energy. For the LEED certification, projects will gain points according to their percentage energy reductions from their baseline energy usage by trying different design alternatives and products. This will take place under the Enhanced Energy Efficiency credit, which contains 10 total points and is the credit that contains the highest number of points.

Following are some strategies that can result in reducing the energy demand of buildings:

- **Building orientation** has the biggest effect on energy demand. By maximizing the southern exposure to the sun, buildings can have increased daylight and passive solar heating. (An example may be to orient the building in order to locate most of the windows facing south.) By minimizing western exposure, annual cooling costs can be dramatically reduced.
- **Community planning** can support building configurations in order to minimize solar gain in summer and maximize it in the winter.
- **Adjacent buildings** can be helpful in providing shade or insulation, or in blocking winds.
- **Passive strategies** like daylighting, thermal mass, and natural ventilation can reduce the energy demand for lighting, heating, and cooling.
- **Efficient building insulation** can reduce energy use.
- **Educating occupants** on their patterns of energy consumption can reduce energy demand.
- **3D computer modeling**, particularly when combined with building information modeling (BIM) software on energy design, can help a lot in evaluating different design alternatives by showing their outcomes. In addition, they can show the best locations for placing windows, doors, and other items.
- In existing buildings, conducting an **energy audit** can highlight the building systems that are not operating efficiently.
- **Rightsizing** is another factor that will affect energy demand. The bigger the building, the higher the energy demand; therefore, projects should not be developed larger than their needs.

USE OF REFRIGERANTS

In addition to reducing energy consumption, green buildings need to address the types of refrigerants used in a project. **Refrigerants** are substances commonly used in air-conditioning and refrigeration systems.

Refrigerants in building systems can deplete stratospheric ozone, a gas that protects human health and the environment by absorbing harmful ultraviolet (UV) radiation. The destructive potential of these chemicals is measured by both their ozone depletion potential (ODP) and global warming potential (GWP). Because hydrochlorofluorocarbon (HCFC) refrigerants are known to have ozone-depleting potential, LEED prohibits their use in new equipment under the Fundamental Refrigerant Management prerequisite, which will be discussed in the following sections.

Refrigerants are most harmful when they leak from cooling and heating systems. If there is any problem with the seals, leaks are likely to occur. For that reason, LEED also requires all projects to check both new and existing refrigerant-containing equipment for leaks.

Natural refrigerants are generally much more environmentally friendly than conventional alternatives. These include carbon dioxide (CO_2), water (H_2O), ammonia (NH_3), air, and hydrocarbons such as propane, ethane, and butane. Their ozone depletion potential (ODP) is zero, and their global warming potential (GWP) is zero or very low. However, their suitability and energy performance vary depending on the refrigerant, the system design, and the application. In some cases, a natural refrigerant system may require more energy than a system using a conventional refrigerant, while in others it can perform very efficiently.

ENERGY EFFICIENCY

After reducing energy demand, the project team should look for ways to improve energy efficiency, which will result in doing the same amount of work with less energy.

ENERGY EFFICIENCY FOR HVAC SYSTEMS

With the integrated process, project teams can employ synergistic strategies. An example of a synergistic strategy may be installing a building envelope that allows for the installation of a smaller HVAC system because of the high level of insulation provided by the envelope.

The U.S. Department of Energy estimated that 40% of the energy used to heat and cool the average building was lost to air leaks in building envelopes. Building envelope problems can lead to high operating costs, uneven building temperatures, poor indoor air quality, building deterioration, noise, and other effects.

An important measure of building envelope performance is the **R-value**, which indicates how well a material resists heat flow; **the higher the R-value, the better the insulation**. However, strong envelope performance also depends on other factors. If insulation is interrupted or poorly installed, heat can still pass through weak points. For this reason, the building envelope must be carefully designed and constructed to minimize air leakage and limit **thermal bridging**, which refers to areas where conductive materials within the envelope permit unwanted heat transfer in cold and hot weather conditions.

A well-performing building envelope must control several things at once: the flow of heat, the flow of air, the flow of water, and the flow of water vapor. This requires insulation, air barriers, waterproofing, and vapor-retarding materials working together as a complete system. If moisture is not properly managed, it can become trapped within the envelope, leading to material damage and mold growth that may affect both occupant health and the durability of the building.

Beyond lowering energy use and improving indoor comfort, a high-performance envelope also strengthens a building's resilience. With a better envelope, indoor conditions can remain safer and more comfortable for a longer period during power outages or extreme weather.

Increasing the thermal mass of a building is another strategy for downsizing HVAC equipment. **Thermal mass** is the ability of a material to absorb and store energy, and in most buildings, a lot of heat is required to change the temperature of high-density materials such as concrete and brick. Using thicker exterior brick or similar exterior building layers will result in lower electricity usage for HVAC, as these materials slowly absorb and store heat during the day while keeping the indoor temperature stable.

In appropriate climates, using **natural ventilation** can be another effective strategy to reduce energy usage and provide fresh air to building occupants. By providing an adequate number of windows and openings, natural ventilation can be achieved.

When choosing HVAC equipment, a project team needs to pay special attention to its operating efficiency. The initial cost of an efficient HVAC system will have a short payback period and will result in energy savings throughout the entire life cycle of the equipment. In the United States, HVAC systems account for 30% of the energy used in commercial buildings and nearly 50% of the energy used in residential buildings.[1]

Another effective strategy is utilizing **energy-recovery ventilation**, in which exhaust air is used to condition incoming fresh air. During cold weather, heat from the outgoing air can be transferred to the incoming air, while in hot weather, the process can work in reverse to reduce the cooling load.

For mechanically ventilated buildings, the air filters also need to be checked regularly, and dust needs to be cleaned. Otherwise, HVAC equipment will need to use more energy to move air through the ducts.

Radiant heat flooring and underfloor air distribution can be other strategies used to address energy efficiency.

Projects can also capture **economies of scale**, and, when possible, multiple buildings can share one HVAC system. By employing district heating and cooling systems, all the shared buildings can take advantage of lower operating costs.

ENERGY EFFICIENCY FOR LIGHTING

For some buildings, lighting can be the biggest contributor to a building's energy use if they are still using incandescent or fluorescent lights. Energy is used both to power the lights and to offset the increased cooling loads resulting from the heat of the lighting. **Daylighting** is surely the best strategy, as it saves energy and increases the well-being of building occupants.

For artificial lighting, **LEDs** have greatly increased energy efficiency and are the clear option that must be used. However, project teams should still ensure that lighting systems provide only the amount of lighting really needed, no more and no less. One important concept in this process is **connected lighting power**, or **lighting power density**, which refers to the amount of power used per square foot or square meter when all lighting is turned on. The appropriate lighting power level depends on the type of space, since the needs of a warehouse are different from those of a hotel lobby. Based on the space type and other relevant factors, the project team can establish a baseline for connected lighting power and then set a target for improvement.

Nevertheless, it is still important to design, commission, and manage lighting systems to provide only the quality and quantity of light needed for the purpose. By installing an adequate number of **lighting controls** for individual users—made up of **photosensors**, **timers**, **occupancy sensors**, and **advanced lighting controls**—projects can cut down their lighting costs since these devices will shut off lights when lighting is not needed.

ENERGY EFFICIENCY FOR APPLIANCES

When choosing appliances for building use, project teams need to consider the energy efficiencies of different alternatives. Appliances with ENERGY STAR labels can be a smart choice since their energy-efficiency criteria have been established by the U.S. Environmental Protection Agency.

ELECTRIFICATION

When fossil fuels are burned, they generate harmful greenhouse gas emissions, in other words, carbon emissions. As a result, electrifying building systems to avoid all on-site combustion of fuels can make an enormous contribution to decarbonization. **Electrification** refers to replacing non-electric energy sources with electricity.

For example, the installation of **heat pumps**, an energy-efficient system for heating and cooling spaces and for heating water, draws thermal energy out of water, air, or the ground into a refrigerant and then transfers that heat to another place. Although they transfer thermal energy, heat pumps work with electricity and can replace conventional fossil-fuel-based systems for space heating and water heating.

Similarly, **electric induction cooking**, which uses a magnetic field to directly heat pots and pans, is much safer than traditional gas systems. It eliminates emissions from on-site combustion, and when the pot is removed, the cooktop stops generating heat. Note that, as this induction technology uses magnets to generate heat, it only works with steel and iron cookware.

As buildings become more electrified, however, power grids will need to handle not only greater overall electricity demand but also new patterns in when that demand occurs. If this shift is not carefully managed, it can place additional stress on grid infrastructure and affect reliability.

Because of this, project teams should focus on creating buildings that are both highly efficient and capable of interacting with the grid in a flexible way. A **grid-interactive building** can adjust how and when it uses electricity depending on grid conditions. This approach helps reduce pressure during periods of high demand, supports the use of renewable energy, and contributes to both decarbonization and long-term resilience.

One important strategy is demand response. Demand response is a mechanism that aims to overcome grid stress by reducing energy demand, especially during peak times. The utility company sends an alert, known as a demand response event or a curtailment event, to customers who agree to adjust their usage patterns during periods of high demand. In turn, consumers reduce their consumption in response to the alert. For example, a commercial building might temporarily delay electric vehicle charging or reschedule certain equipment to operate later in the day. Small adjustments like these, when made across many buildings, can significantly reduce peak demand and support grid stability.

Another useful strategy is energy storage at the project site. **Batteries** can store electricity either from on-site renewable systems or from the grid during hours when demand is lower, and that stored electricity can then be used later when demand rises.

Thermal energy storage takes a different approach. Instead of storing electricity, it stores heating or cooling capacity for later use, typically in an insulated tank. One of the most effective methods is **ice storage**, which takes advantage of the large amount of energy water absorbs and releases during freezing and thawing. A building can produce ice during off-peak hours and then use it to cool the building during the day when demand is highest. This type of system is sometimes referred to as a thermal battery, even though it stores heat rather than electricity.

Electrification alone does not guarantee decarbonization. For the built environment to become truly low-carbon, the electricity used by buildings must also come increasingly from renewable energy sources, which is our next topic.

RENEWABLE ENERGY

As the name implies, renewable energy is a type of energy that derives from renewable sources. **Renewable energy** includes solar, wind, wave, biomass, and geothermal power, plus certain forms of hydropower. The use of renewable energy sources avoids the environmental impacts associated with the production and consumption of nonrenewable fuels, such as coal, nuclear power, oil, and natural gas.

Renewable energy sources can be either on-site or off-site.

On-site renewable energy production involves systems that generate renewable energy, such as solar photovoltaic (PV) panels, which convert sunlight into electricity. **Solar thermal collectors** can also capture solar energy to heat air or water for uses such as space heating, water heating, and swimming pool heating.

Off-site renewable energy production can involve the following sources:

- **Wind energy:** Wind turbines convert the kinetic energy of moving air into electricity.
- **Hydropower:** Hydropower converts the energy of flowing or falling water into electricity.
- **Bioenergy:** Electricity can be generated from biomass- or biofuel-based resources such as wood waste, landfill gas, agricultural crops or residues, animal waste, and other organic materials.
- **Wave power:** Wave energy converters can convert the energy of ocean waves into electricity.
- **Tidal power:** Tidal power is a form of hydropower that converts the energy of tides into electricity.

Wind Turbines *(Photo courtesy of Andy Dingley)*

Off-site renewable energy is procured through contracts with utilities or the purchase of renewable energy certificates (RECs). If the local utility provider sells renewable electricity, the project can purchase it directly. If not, project teams can buy **renewable energy certificates (RECs)**, which are tradable, intangible instruments representing the environmental and other non-power attributes of renewable electricity generation. One REC generally represents 1 megawatt-hour (MWh) of electricity generated from an eligible renewable resource and delivered to the grid. Purchasing RECs can help support renewable energy development by providing an additional source of revenue for renewable energy generators. (RECs will be discussed further under the Renewable Energy credit.)

POWER RESILIENCE

Project teams should also consider what will happen if the grid fails during an emergency. Power outages often occur during severe weather events or other disruptions, when access to electricity may be especially important for safety and communication. For this reason, projects should plan for a reliable backup power strategy.

Renewable backup power can offer significant advantages over conventional fossil-fuel generators. Systems such as solar panels or wind turbines are not dependent on fuel deliveries and can continue producing energy during longer outages if properly designed. To be dependable, however, these systems must be paired with controls and usually with energy storage so that electricity can be managed effectively when the grid is unavailable.

Projects with on-site renewable energy should also address how the system will operate when the grid goes down. A grid-connected renewable energy system cannot automatically continue supplying power during an outage, because sending electricity back into a damaged or inactive grid can create serious safety risks, especially if workers are repairing infrastructure. To avoid this, the system must be able to disconnect from the main grid and operate independently. This capability, known as **islanding**, allows the building to safely use its on-site energy resources during an outage and continue providing essential services even when the wider grid is not functioning.

BUILDING COMMISSIONING AND PERFORMANCE VERIFICATION

With the Fundamental Commissioning prerequisite, LEED requires all projects to implement building commissioning. **Commissioning** is a systematic process carried out by skilled personnel to verify that a building's systems are designed, installed, and operating in accordance with the owner's project requirements (OPR), project goals, and design specifications. Building commissioning ensures that the project is designed and built as truly intended.

The process begins early in design by first specifying the OPR. These requirements are then incorporated into the building design and implemented during construction. However, a green building is not completed at the end of the construction phase. During the building operations phase, it is critical to ensure that a project functions as designed and sustains its performance over time.

As performance goals established during planning and design may be undermined by construction defects or equipment malfunctions, building systems should be monitored during building operations. Two key strategies help ensure that building systems continue to function as designed: monitoring-based commissioning (MBCx) and retrocommissioning.

Monitoring-based commissioning (MBCx) is the process of using software to monitor real-time data from building automation systems and meters, enabling building operators to detect faulty equipment, identify unusual energy consumption patterns, and address issues as they occur. The work is led by a monitoring-based commissioning provider (MBCxP)—either a contracted third party or a qualified in-house staff person.

Retrocommissioning, on the other hand, is a systematic process for analyzing and optimizing building system performance involving a commissioning provider. It includes testing of building systems and documentation of any measures needed to improve performance. Retrocommissioning is typically applied to existing buildings to bring them back on track for meeting or exceeding their original operational goals.

In summary, retrocommissioning is a point-in-time evaluation and optimization effort, while MBCx is a continuous, software-based monitoring approach that keeps performance on track over time.

Even though the commissioning process results in extra costs to the owner, commissioning is one of the most cost-effective solutions for improving building performance and cutting down operating costs.

STRATEGIES TO ADDRESS ENERGY AND ATMOSPHERE

- Increase energy efficiency by calculating the building's baseline energy use and identifying innovative ways to reduce overall energy consumption.
- Size the building appropriately to meet project needs and avoid overdevelopment.
- Orient the building to minimize energy use.
- Create a high-performance building envelope designed and constructed to minimize air leakage, limit thermal bridging, and prevent mold growth.
- Use a high-efficiency HVAC system.
- Utilize energy-recovery ventilation.
- Use natural ventilation where appropriate.
- Use high-efficiency appliances.
- Use energy-efficient lighting.
- Install lighting controls, photosensors, timers, occupancy sensors, and advanced lighting controls.
- Electrify building systems to avoid on-site combustion.
- Prefer heat pumps over conventional fossil-fuel-based systems for space heating and water heating.
- Prefer electric induction cooking over traditional gas systems.
- Implement grid-interactive building strategies.
- Participate in demand response programs.
- Use no refrigerants or use refrigerants with low ODP and GWP values.
- Check both new and existing refrigerant-containing equipment for leaks.
- Generate on-site renewable energy.
- Store renewable energy on site using batteries.
- If installing on-site renewable energy is not possible, purchase off-site renewable energy from the local utility provider or purchase renewable energy certificates (RECs).
- Implement building commissioning during the design and construction phases.
- Implement monitoring-based commissioning (MBCx) and retrocommissioning during the building operations phase.

ENERGY AND ATMOSPHERE PREREQUISITES AND CREDITS

OPERATIONAL CARBON PROJECTION AND DECARBONIZATION PLAN—PREREQUISITE

PREREQUISITE SUMMARY

This prerequisite is designed to make project teams and building owners think about energy and carbon during the early stages of the design process. Instead of waiting until late design to "value engineer" sustainability out of the project, LEED asks teams to evaluate efficiency, peak load reduction, and decarbonization measures during the early stages.

The prerequisite requires the project team to estimate the building's annual site energy use by energy type and submit it to USGBC. Using that energy estimate, the project's location, and current grid data, USGBC then creates a business-as-usual (BAU) carbon projection showing what the building's emissions from energy use could look like over the next 25 years.

Finally, the project must address how it will move toward decarbonization over that same 25-year period. This can be done by demonstrating strong electrification performance through the Electrification credit, or by preparing a short decarbonization plan that lays out the likely future retrofit steps, and their timelines and costs. In other words, the goal is to show the project team and the building owner how much money can be saved by designing the building for decarbonization now, rather than paying for major retrofits later—such as envelope upgrades and the replacement of relatively new, expensive heating equipment.

PREREQUISITE REQUIREMENTS

To comply with this prerequisite, the project must complete all of the following requirements.

1) Design Analysis:

Early in design, analyze strategies related to energy efficiency, peak load reduction, and decarbonization, then use the results to influence design decisions.

The analysis must use at least one of the following methods:

- Simplified energy modeling
- Analysis from similar projects
- Analysis from published data

AND

2) Site Energy Estimate:

Estimate the project's annual site energy use by energy type and submit the results to USGBC.

AND

3) Review Carbon Projection:

Using the submitted annual energy data, the project's location, and current grid data, USGBC will create a "business as usual" (BAU) projection of the project's carbon emissions from energy use over a 25-year period. This projection will inform owners and designers on the carbon impact of their decisions over the next 25 years.

AND

4) Decarbonization Plan:

Projects that have been substantially electrified by achieving four or five points in the Electrification credit are exempt from this requirement, as they have already been designed for the low-carbon future. Other projects, on the other hand, should prepare a decarbonization plan that explains how decarbonization could be achieved over a 25-year period through future retrofits.

KEY THINGS TO REMEMBER

- Evaluating efficiency, peak load reduction, and decarbonization measures during the early stages of design can create impactful and cost-effective strategies that can integrate into the project.
- With this prerequisite, LEED requires projects that have not been substantially electrified to create a decarbonization plan to show the project team and the building owner how much money can be saved by designing the building for decarbonization now, rather than paying for major retrofits later.

MINIMUM ENERGY EFFICIENCY—PREREQUISITE

PREREQUISITE SUMMARY

Just as the Minimum Water Efficiency prerequisite aims to decrease potable water usage of a building, this prerequisite aims to establish a minimum energy performance. It establishes the minimum threshold of energy efficiency that every LEED project must meet, serving as the foundation upon which all other Energy and Atmosphere credits build. The prerequisite requires compliance with ANSI/ASHRAE/IES Standard 90.1, the industry's benchmark for energy-efficient design.

ANSI/ASHRAE/IES Standard 90.1 traditionally evaluates building performance using energy cost—comparing the proposed building's annual energy cost against a baseline. However, cost does not always reflect environmental impact, because cheap fossil fuels can score well on cost while producing significant carbon emissions. To address this, LEED allows projects to use an alternative metric called future source energy. This metric looks beyond what happens at the building and considers the total energy needed to generate and deliver that energy, including losses along the way.

The prerequisite also recognizes that energy performance expectations must vary by climate and building type. A building in a hot, humid climate zone faces fundamentally different energy challenges than a building in a cold climate. To account for this, ANSI/ASHRAE/IES Standard 90.1 assigns Building Performance Factors (BPFs) that set different performance targets for each combination of building type and climate zone.

PREREQUISITE REQUIREMENTS

Comply with **ANSI/ASHRAE/IES Standard 90.1**. Projects registering before January 1, 2028, may use either the 2019 edition or 2022 edition. Projects registering on or after January 1, 2028, must use the 2022 edition.

For projects applying the Performance Rating Method in Normative Appendix G, future source energy may be used in place of cost as the performance metric.

KEY THINGS TO REMEMBER

- LEED requires every project to meet a minimum energy performance baseline by complying with ANSI/ASHRAE/IES Standard 90.1.

FUNDAMENTAL COMMISSIONING—PREREQUISITE

PREREQUISITE SUMMARY

As a major part of the integrated process, building commissioning ensures that projects are designed and built as truly intended. A well-oriented commissioning process will always lead to fewer change orders, avoidance of project delays and cost overruns, and increased durability of the whole project.

This prerequisite forces projects to implement a fundamental commissioning process to ensure that the project design and construction meet the owner's project requirements (OPR) and basis of design (BOD). Project teams will additionally develop an ongoing commissioning plan to maintain system performance throughout the building's operational life.

By the end of the design development phase, a qualified commissioning provider (CxP) must be designated. The CxP must have direct commissioning experience on at least two projects of equal or larger scope and complexity and must be independent of the design and construction team, reporting directly to the owner. The CxP can be a third-party consultant with no involvement in the project's design or construction, an employee of the owner provided they meet the qualification requirements, or an employee of the design or construction firm as long as they are completely independent from any aspect of the project's design or installation. If the CxP employs any company with direct influence on the design and construction, the commissioning plan must clearly address potential conflicts of interest and demonstrate that the CxP acts solely on behalf of the owner and works entirely independently from the design and construction team.

The CxP will assist in developing the OPR, review the BOD for compliance with the OPR, develop and implement a commissioning (Cx) plan, confirm incorporation of Cx requirements into the construction documents, develop functional performance tests, verify system performance, and prepare a final Cx report while documenting all findings directly to the owner throughout the whole process.

The Cx Plan will contain the following:

- Goals and objectives
- General project information
- Systems to be commissioned
- Team member roles and responsibilities
- Key tasks and milestones
- OPR and BOD reviews
- Functional performance testing procedures
- Verification of system performance
- Reporting deficiencies and resolving issues

The Cx plan will be a living document that is updated throughout the project and will also form the basis of the final Cx report. To ensure designers and engineers are meeting the OPR and BOD, the CxP will attend at least one coordination meeting focused on mechanical, electrical, and plumbing systems and one focused on the building envelope, which may be combined into a single meeting.

During construction, the CxP will review submittals and substitutions for design deviations that impact the OPR, attend milestone meetings at 50% and 100% completion, and perform a sample review of at least 10% of completed contractor documentation for quality assurance. The CxP will witness functional performance testing among the building elements and systems to make sure everything is installed according to the specifications and the design. For the building envelope, testing must be included in the commissioning documents.

All the data used during the design and construction phase will be compiled by the CxP into a final Cx report, which will contain a Cx process overview, OPR, BOD, design review logs, approved submittals, Cx specifications, functional performance tests, and an issues and resolution log identifying closed items and proposed resolutions for outstanding items.

Finally, the CxP should develop an ongoing commissioning plan to ensure systems remain operationally efficient throughout the building's life. The plan should provide facility managers with procedures, blank functional performance tests, and a recommended schedule for ongoing Cx activities.

Systems that are part of the fundamental commissioning process include:

- HVAC systems
- Service water heating
- Power systems
- Lighting and controls
- On-site renewable energy systems
- Energy monitoring systems
- Refrigeration equipment
- Energy storage systems
- Load management systems
- Building envelope
- Other energy-using systems within the project boundary

PREREQUISITE REQUIREMENTS

Comply with the commissioning requirements of **ANSI/ASHRAE/IES Standard 90.1** for building systems, controls, and the building envelope. Projects registered before January 1, 2028, must reference Standard 90.1-2019 or later; projects registered on or after that date must use Standard 90.1-2022 or later.

By the end of the design development phase, the owner must designate a commissioning provider (CxP) with direct experience completing commissioning on at least two projects of equal or larger scope and complexity.

In predesign or as early as possible, the CxP must assist in developing the owner's project requirements (OPR), which must address HVAC, service water heating, power, lighting, on-site renewable energy, other equipment, and the building envelope. The CxP must also review the basis of design (BOD) for compliance with the OPR and attend at least one coordination meeting focused on mechanical, electrical, and plumbing systems and one focused on the building envelope, which may be combined.

During construction, the CxP reviews submittals for design deviations impacting the OPR, attends milestone meetings at 50% and 100% completion, monitors functional performance testing, and performs a sample review of at least 10% of contractor documentation for quality assurance. For the building envelope, testing must be included in the commissioning documents with CxP witnessing of a sample of tests.

Following occupancy, the CxP must develop an ongoing commissioning plan to maintain system performance over the building's operational life.

KEY THINGS TO REMEMBER

- With this prerequisite, LEED forces every project to undergo a structured commissioning process to verify that building systems are designed, installed, and operating in accordance with the owner's project requirements.
- LEED refers to ANSI/ASHRAE/IES Standard 90.1 for fundamental commissioning.
- The CxP can be a third-party consultant with no involvement in the project's design or construction, an employee of the owner provided they meet the qualification requirements, or an employee of the design or construction firm as long as they are completely independent from any aspect of the project's design or installation.
- The CxP must always report directly to the owner.
- The CxP must have direct commissioning experience on at least two projects of equal or larger scope and complexity.
- The Cx plan is a living document updated throughout the project and forms the basis of the final Cx report.
- The final Cx report must include a process overview, OPR, BOD, design review logs, approved submittals, functional performance tests, and an issues and resolution log.
- LEED requires developing an ongoing commissioning plan to maintain system performance after occupancy.

ENERGY METERING AND REPORTING–PREREQUISITE

PREREQUISITE SUMMARY

Similar to the Water Metering and Reporting prerequisite, which tracks and reports building water use, this prerequisite requires the tracking and reporting of building energy use and demand.

All projects should commit to sharing their monthly energy data with USGBC, including total energy consumption for each energy source, on-site renewable energy generation, and peak electrical demand.

PREREQUISITE REQUIREMENTS

Project teams should comply with the following requirements:

1) Energy Monitoring and Recording:

Install (or use existing) devices to monitor and record energy use per **ANSI/ASHRAE/IES Standard 90.1**.

The required version of Standard 90.1 depends on the project registration date:

- 2019 or later for projects registered before January 1, 2028.
- 2022 or later for projects registered on or after January 1, 2028.

Additionally, install (or use existing) devices to monitor and record energy use for on-site renewable electricity generation.

AND

2) Report Energy Data:

Commit to reporting monthly energy data to USGBC at least once a year.

The report must include 12 consecutive months of total energy use by source, on-site renewable energy generation, and peak electrical demand.

Continue reporting for five years, or until the building changes ownership or lessee.

KEY THINGS TO REMEMBER

- LEED references ANSI/ASHRAE/IES Standard 90.1 for energy metering.
- LEED requires projects to track both building energy use by source and on-site renewable energy generation.
- LEED requires all projects to report their monthly energy data to USGBC at least once a year, including total energy use by source, on-site renewable energy generation, and peak electrical demand.

FUNDAMENTAL REFRIGERANT MANAGEMENT—PREREQUISITE

PREREQUISITE SUMMARY

Hydrochlorofluorocarbon (HCFC) and chlorofluorocarbon (CFC) refrigerants carry extremely high global warming potentials (GWPs) and have caused serious damage to the ozone layer. The Montreal Protocol, signed in 1987, targeted the most harmful ozone-depleting substances, and CFC production was phased out in participating countries before 1995. Later, the Kigali Amendment to the Montreal Protocol expanded these efforts by driving the phaseout of HCFCs in most developed countries. However, these refrigerants can still be found in some developing regions. Under this prerequisite, LEED prohibits the use of HCFC refrigerants in any new equipment.

Refrigerants are most harmful when they leak from cooling and heating systems. If there is any problem with the seals, leaks are likely. For that reason, this prerequisite also requires all projects to check both new and existing refrigerant-containing equipment for leaks.

Projects that do not use any refrigerants can earn this prerequisite without further action under Option 1.

PREREQUISITE REQUIREMENTS

Option 1: No Refrigerants

Do not use any refrigerants in the project.

OR

Option 2: Refrigerants

Project teams should meet the following requirements:

- Create an inventory of all refrigerant-containing equipment installed within the project scope of work, plus any existing refrigerant-containing equipment owned by the building owner. The inventory must list the refrigerant type, global warming potential (GWP), the amount of refrigerant in each piece of equipment, and the total GWP of all refrigerants. Equipment that contains less than 0.5 pounds (225 grams) of refrigerant can be excluded from the calculations.
- Do not use HCFC refrigerants in any new equipment.
- During design, evaluate available alternatives for any refrigerant with GWP greater than 700.
- Before substantial completion, check both new and existing refrigerant-containing equipment for leaks and fix any leaks found. For systems with field-assembled joints, perform a leak check, vacuum check, and pressure check before charging the system with refrigerant.

KEY THINGS TO REMEMBER

- Project teams should select cooling and heating systems that use refrigerants with the lowest global warming potential (GWP) possible and design systems to use no more refrigerant than necessary. Additionally, project teams should implement robust leak-prevention and detection measures to ensure refrigerants remain in the systems. At the end of such equipment's service life, the refrigerants should be captured and then either recycled or destroyed.
- Equipment that contains less than 0.5 pounds (225 grams) of refrigerant is exempt from LEED requirements.

ELECTRIFICATION—CREDIT

- 1–5 Points (New Construction)

CREDIT SUMMARY

On-site combustion of fossil fuels for heating, hot water, and cooking is one of the largest sources of direct carbon emissions from buildings.

This credit targets those emissions by rewarding projects that eliminate or significantly reduce their reliance on combustion-based systems in favor of efficient electric alternatives such as heat pumps.

The shift away from on-site combustion is a cornerstone of building decarbonization, reducing both direct emissions and the health risks associated with indoor fuel burning.

CREDIT REQUIREMENTS

Option 1: No On-Site Combustion (5 points)

Project teams should design and operate the project with no on-site combustion from start-up, except for emergency support systems.

The combined weighted average equipment efficiency for space heating and service water heating should be at least 1.8 COP. COP, or coefficient of performance, measures how efficiently a system converts energy into heating or cooling; a COP of 1.8 means the system delivers 1.8 units of heat for every 1 unit of energy consumed.

OR

Option 2: No On-Site Combustion Except at Low Temperatures (1–4 points)

Project teams should pursue any combination of the following paths.

Path 1 eliminates combustion for space heating except at low temperatures. Projects in climate zones 3 and above—zones with cooler winters where heating demand is significant—must achieve a weighted average equipment efficiency of at least 1.8 COP.

Path 2 applies the same approach to service water heating, requiring either a minimum 1.8 COP or a solar hot water fraction of at least 0.4, meaning solar energy covers at least 40% of the hot water load.

Path 3 eliminates combustion for cooking, laundry, and other process loads—projects without these systems earn this point automatically.

KEY THINGS TO REMEMBER

- On-site combustion of fossil fuels for heating, hot water, and cooking is one of the largest sources of direct carbon emissions from buildings. LEED rewards decarbonization achieved through electrifying building systems traditionally fueled with on-site combustion, including space heating, service water heating, cooking, and other process equipment.

REDUCE PEAK THERMAL LOADS—CREDIT

- 1–5 Points (New Construction)

CREDIT SUMMARY

The building envelope is the primary physical barrier between a building's conditioned interior and the outside environment. How well that envelope performs—how much air it leaks, how effectively it manages heat transfer, and how efficiently it works with mechanical systems—has a direct impact on a building's energy consumption and carbon emissions. This credit targets envelope performance by rewarding projects that go beyond baseline code requirements to reduce infiltration, recover ventilation energy, address thermal bridging, and lower peak heating and cooling loads.

The credit offers four options that can be pursued individually or in combination for up to 5 points. Option 1 focuses on airtightness and balanced ventilation, requiring measured air leakage testing and airflow balancing. Option 2 requires energy or heat recovery on ventilation systems. Option 3 addresses thermal bridging, which occurs when highly conductive materials in the envelope create pathways for heat to bypass insulation. Option 4 rewards projects that demonstrate measurable reductions in peak thermal loads through Passive House tools, ASHRAE 90.1 trade-off methods, or energy simulation. Together, these options push project teams to treat the envelope as an integrated system rather than a collection of individual components.

By reducing peak heating and cooling demands at the envelope level, projects can downsize mechanical equipment, lower operational energy use, and improve occupant comfort.

CREDIT REQUIREMENTS

Projects may comply with any combination of Options 1 through 4 for a maximum of 5 points.

Option 1: Infiltration and Balanced Ventilation (1 point)

Design ventilation and exhaust airflows within 10% of each other and include a testing, adjusting, and balancing (TAB) report in the commissioning scope. Additionally, conduct an air leakage test to demonstrate that the measured air leakage of the building envelope is at or below the thresholds defined by LEED.

AND/OR

Option 2: Ventilation Energy Recovery (1 point)

Each fan system supplying outdoor air must include an energy or heat recovery system with provisions to bypass recovery during moderate outside air conditions.

AND/OR

Option 3: Thermal Bridging (1 point)

Projects must demonstrate compliance with the thermal bridging requirements of ASHRAE 90.1-2022.

AND/OR

Option 4: Peak Thermal Load Reductions (1–3 points)

Reduce the peak heating and cooling demands placed on building systems. Measure building envelope air leakage and demonstrate balanced ventilation, then pursue one of three calculation paths. Path 1 caps the combined peak heating and cooling load per unit of floor area using Passive House calculation tools. Alternatively, Path 2 demonstrates a percentage improvement in peak loads by comparing proposed envelope and mechanical performance against ASHRAE 90.1 baseline values. Path 3 uses whole-building energy simulation per ASHRAE 90.1 Appendix G to show reductions in peak coincident loads.

KEY THINGS TO REMEMBER

- How well the building envelope performs directly impacts a building's energy consumption and carbon emissions. By tightening infiltration, installing heat or energy recovery on ventilation systems, minimizing thermal bridging, and reducing peak heating and cooling loads, projects can downsize mechanical equipment, lower energy use, and improve occupant comfort.

ENHANCED ENERGY EFFICIENCY—CREDIT

- 1–10 Points (New Construction)

CREDIT SUMMARY

This credit builds on the Minimum Energy Efficiency prerequisite and awards points for additional energy reductions. It is the highest-value credit in the Energy and Atmosphere category, offering up to 10 points for projects that demonstrate significant energy performance improvements through either prescriptive measures or whole-building energy simulation.

The credit offers two options. Option 1, the Prescriptive Path, allows teams to earn points by implementing specific, predefined efficiency measures without running a full energy model. It has two paths that can be combined. Path 1 addresses regulated loads by implementing additional efficiency measures from ANSI/ASHRAE/IES Standard 90.1, Section 11, beyond what the code requires. Path 2 addresses plug and process loads. Teams can earn points through plug load management strategies, by installing ENERGY STAR or equivalently efficient equipment across multiple categories, or by demonstrating percentage improvements in total plug and process loads.

Note that **nonprocess**, or **regulated energy** includes items that are used to condition spaces and maintain comfort and amenities for building occupants such as:

- Interior and exterior lighting (parking garage, security, landscape, architectural lighting)
- HVAC (heating, cooling, fans, pumps)
- Hot water heating
- Toilet exhaust
- Parking garage ventilation

Process energy or **unregulated energy**, on the other hand, includes:

- Computers
- Office equipment
- Kitchen stoves
- Kitchen refrigerators
- Cooking and food preparation equipment
- Washers and dryers
- Elevators and escalators

The energy demands created by nonprocess energy are referred to as **nonprocess loads**, while those created by process energy are referred to as **process loads**.

Note that most process energy items are connected to the building's electrical system through standard receptacles, with the exception of elevators and escalators. Because of this, the term "process load" is often synonymously used with **plug loads** (**receptacle loads**), which represent the electrical use by all the equipment that is connected to the electrical system via electrical receptacle. Strictly speaking, however, plug loads are only a subset of process loads, because not all process loads are connected through a receptacle.

While Option 1 breaks energy performance into individual prescriptive measures, Option 2, Energy Simulation, takes a whole-building approach. Instead of checking off individual measures, the project team builds an energy model comparing the proposed building against a baseline. Performance is measured using the future source energy metric. Points are awarded based on the percentage improvement in the Performance Index (PI) below a target value, with greater reductions earning more points. Projects can calculate their improvement either excluding or including the contribution of on-site renewable energy systems, with the path that includes renewables requiring a larger percentage reduction to earn the same number of points.

CREDIT REQUIREMENTS

Option 1: Prescriptive Path (1–10 points)

Project teams should pursue Path 1 and/or Path 2 for up to 10 points.

Projects using ASHRAE 90.1-2019 (available only for registrations before January 1, 2028) earn points by implementing additional efficiency measures from ASHRAE 90.1-2022, Section 11, with points awarded based on the number of credits achieved—starting at 1 point for 25 credits up to 5 points for 125 credits. Projects using ASHRAE 90.1-2022, on the other hand, must first meet the minimum Section 11 credits required by that standard and then earn points by implementing incremental credits beyond that minimum.

Path 2 addresses plug and process loads through one of three cases. Case 1 earns 1 point for implementing plug load management, including an occupant-accessible plug load dashboard and policies for powering down IT equipment when not in use. Case 2 earns 1 to 4 points for installing ENERGY STAR-rated or equivalently efficient equipment across categories such as office equipment, appliances, commercial food service, data centers, elevators, and refrigeration systems—with 90% of applicable equipment by quantity or rated load meeting the criteria. Case 3 earns 1 to 4 points by demonstrating a percentage improvement in total plug and process loads using the ASHRAE 90.1 Exceptional Calculation Method, with 10% earning 1 point up to 40% earning 4 points.

OR

Option 2: Energy Simulation (1–10 points)

Demonstrate an improvement in future source energy using the ASHRAE 90.1 Performance Rating Method (Appendix G). The proposed building is modeled against a baseline building and performance is expressed as a Performance Index (PI). Points are awarded based on the percentage reduction of the PI below a target value. Projects may calculate improvement either excluding on-site renewable contribution, where a 3% reduction earns 1 point and a 30% reduction earns 10 points, or including on-site renewable contribution, where a 10% reduction earns 1 point and a 100% reduction earns 10 points. Path 2 requires larger percentage improvements because it accounts for the added benefit of on-site renewable generation. The Building Performance Factors are adjusted by building type and climate zone.

KEY THINGS TO REMEMBER

- Nonprocess energy, also called regulated energy, includes building systems used to condition spaces and provide comfort and basic amenities for occupants, such as interior and exterior lighting, HVAC systems, hot water heating, toilet exhaust, and parking garage ventilation.
- Process energy, also called unregulated energy, includes equipment tied to building use and operations rather than occupant comfort, such as computers, office equipment, kitchen stoves and refrigerators, cooking and food preparation equipment, washers and dryers, and elevators and escalators.
- Plug loads (receptacle loads), which represent the electrical use of all equipment connected to the electrical system through electrical receptacles, are a subset of process loads because not all process loads are connected through a receptacle.

RENEWABLE ENERGY—CREDIT

- 1–5 Points (New Construction)

CREDIT SUMMARY

Using energy from renewable sources benefits both the environment and human health by reducing carbon emissions and air pollution. This credit awards projects that produce renewable energy on-site or procure Green-e Energy certified renewable energy. **Green-e** is the leading certification program for renewable energy generation in the United States.

Because producing renewable energy on-site eliminates the losses that occur during electricity transmission and delivers greater environmental value than procuring renewable energy, the credit awards more points for on-site generation at lower thresholds than for off-site procurement. LEED structures this distinction through a three-tier system that awards progressively more credit for renewables with a closer, more direct connection to the project. Projects can also earn points through a combination of on-site and off-site sources.

When a renewable energy facility generates electricity, it also produces environmental attributes that represent the clean energy benefits of that generation. These attributes are tracked through tradable instruments called renewable energy certificates (RECs) or energy attribute certificates (EACs). (EAC is the broader international term used globally, while REC is the term most commonly used in the U.S.)

A single **renewable energy certificate** represents the environmental benefits associated with one megawatt-hour of renewable electricity generation. Because RECs can be bought and sold separately from the physical electricity, LEED requires that all RECs or EACs associated with the project's renewable energy be retired on behalf of the LEED project. Retiring a certificate means permanently claiming its environmental benefits so that no other party can use the same certificate to make a renewable energy claim. This prevents double counting.

CREDIT REQUIREMENTS

Supply or procure renewable energy from one or more of three tiers. Tier 1 covers on-site renewable energy generation, including systems located on the project site, on the surrounding campus, or on the site of a social impact project where the system is donated and ownership is transferred to the recipient. Eligible renewable resources include solar electric (photovoltaic), solar thermal, wind, and/or recovered heat from municipal wastewater. Tier 2 covers new off-site renewable electricity from generation assets contracted to be operational within two years of occupancy or within five years of their commercial operations date. Tier 3 covers off-site renewable electricity or fuels that are Green-e Energy certified or equivalent. Points from multiple tiers may be combined, and higher percentages of annual site energy covered by renewables earn more points.

Renewable energy contracts must be at least 10 years in length, or prorated across 10 years for shorter terms. All environmental attributes—energy attribute certificates (EACs) or renewable energy certificates (RECs)—must be retired on behalf of the LEED project.

KEY THINGS TO REMEMBER

- When a renewable energy facility generates electricity, it also produces environmental attributes that represent the clean energy benefits of that generation. These attributes are tracked through tradable instruments called energy attribute certificates (EACs) or renewable energy certificates (RECs).
- A single renewable energy certificate represents the environmental benefits associated with one megawatt-hour of renewable electricity generation.
- Because RECs can be bought and sold separately from the physical electricity, LEED requires that all RECs or EACs associated with the project's renewable energy be retired on behalf of the LEED project, which prevents double counting.

ENHANCED COMMISSIONING—CREDIT

- 1–4 Points (New Construction)

CREDIT SUMMARY

This is the credit of the Fundamental Commissioning prerequisite, which rewards projects that take commissioning a step further. Enhanced commissioning deepens the commissioning provider's (CxP) involvement from predesign through post-occupancy, expands the scope of systems verified, and adds monitoring tools to verify ongoing operational performance.

Option 1 contains two paths in which the first path focuses on MEP systems and requires the CxP to be engaged during predesign or very early in the design phase—earlier than the prerequisite requires. The CxP must comply with ASHRAE Standard 202-2024, which provides a more comprehensive commissioning framework covering mechanical, electrical, plumbing, control, data center, process, building monitoring, and renewable energy systems. Beyond the fundamental requirements, the CxP must attend at least two design phase coordination meetings instead of one and must review training materials and confirm that facility staff training has occurred before or during occupancy. The second path focuses on the building enclosure (also referred to as the building envelope) and requires compliance with ASTM E2947-21a, a standard guide specifically for enclosure commissioning. The enclosure must undergo field testing for air leakage, water penetration, and infrared imaging where applicable. A building enclosure commissioning provider (BECxP) leads this process and can be the same entity as the CxP or a separate one.

In Option 2, on the other hand, project teams implement monitoring-based commissioning, which shifts the focus from construction-phase verification to ongoing operational performance. As mentioned, monitoring-based commissioning (MBCx) is the process of using software to monitor real-time data from building automation systems and meters, enabling building operators to detect faulty equipment, identify unusual energy consumption patterns, and address issues as they occur. The work is led by a monitoring-based commissioning provider (MBCxP)—either a contracted third party or a qualified in-house staff person—who coordinates with the CxP and facilities management team.

CREDIT REQUIREMENTS

Option 1: Enhanced Commissioning (1–3 points)

The owner must designate an independent CxP during predesign or very early in the design phase and choose one or a combination of the two paths.

Path 1 (2 points) requires compliance with **ASHRAE Standard 202-2024** for mechanical, electrical, plumbing, control, data center, process, building monitoring, and renewable energy systems. The CxP must attend at least two design phase coordination meetings and must review training materials and confirm that facility staff training occurred prior to or during occupancy.

Path 2 (1 point), on the other hand, requires compliance with **ASTM E2947-21a** for building enclosure commissioning, including field testing for air leakage, water penetration, and infrared imaging where applicable. The CxP and the building enclosure commissioning provider (BECxP) can be the same entity or separate entities.

AND/OR

Option 2: Monitoring-Based Commissioning (1–2 points)

Project teams should pursue Path 1 or Path 2.

Path 1, Basic Software (1 point), requires a minimum three-year commitment to MBCx through a contracted provider or qualified in-house staff, starting no later than building occupancy. MBCx must be fully coordinated between the commissioning provider, facilities management, and MBCxP.

The team must develop an MBCx plan covering roles, software tools, and reporting protocols—including staff training, prompt communication of faults, quarterly issue summaries, annual performance reports, and at least two MBCxP reviews of building systems and controls.

Additionally, provide a remotely accessible energy information system that tracks and visualizes all metered energy data referenced in the Energy Metering and Reporting prerequisite. The system must allow building operators to benchmark energy use, compare consumption across different time intervals, analyze hourly load patterns, and view submetered electric data. Major electric loads—such as elevators, commercial kitchen equipment, and process equipment in spaces exceeding 10 kW—must also be monitored and visualized on an hourly basis.

Path 2, Enhanced Software (2 points), requires compliance with all Path 1 requirements plus additional monitoring capabilities. Projects with large HVAC or refrigeration systems must provide a fault detection and diagnostics (FDD) system that covers at least 60% by capacity of air handling and large hydronic or refrigeration equipment. The FDD system must identify and visualize faults, link them to relevant trend data, support filtering and reporting, and retain data for at least three years.

The energy information system must also be expanded to include automated detection of energy use anomalies, energy consumption normalization, and greenhouse gas emissions reporting. For major renovations and buildings less than 25,000 sq. ft., teams must comply with ASHRAE 90.1, Section 8.4.3 metering requirements and provide automated reporting of energy anomalies.

KEY THINGS TO REMEMBER

- Involving the CxP as early as possible has many advantages, which is why the credit requires the CxP to be engaged during predesign or very early in the design phase.
- LEED refers to ASHRAE Standard 202-2024 for enhanced commissioning of mechanical, electrical, plumbing, control, data center, process, building monitoring, and renewable energy systems.
- LEED refers to ASTM E2947-21a for building enclosure commissioning, which includes field testing for air leakage, water penetration, and infrared imaging where applicable.

- A building enclosure commissioning provider (BECxP) leads the enclosure commissioning process and can be the same entity as the CxP or a separate one.
- Monitoring-based commissioning (MBCx) uses software to monitor real-time data from building systems and meters, allowing operators to detect faults and unusual energy patterns as they occur. The process is led by a monitoring-based commissioning provider (MBCxP).

GRID INTERACTIVE—CREDIT

- 1–2 Points (New Construction)

CREDIT SUMMARY

Think about a power plant that serves the whole city and imagine that the weather gets extremely hot beyond the normal average temperature. As a consequence, everyone simultaneously turns on his or her air conditioner, creating a sudden increase in electricity demand.

If the power plant's energy is not enough to handle that peak demand, then the utility company would think about constructing an additional plant, or it would need to find additional generation sources, including nonrenewable energy sources.

Demand response is a technology that aims to overcome these types of conflicts by reducing energy demand, especially during peak times. The utility company sends an alert, which is called a DR event or a curtailment event, to customers who agree to change their usage patterns at peak demands. In turn, consumers reduce their demand with the alert. Consumers are rewarded for their participation in the demand response programs, and the construction of additional power plants is avoided. In some cities, utility companies may charge extra during peak times in order to reduce energy demand.

On top of this, demand response programs are helpful in balancing the supply of renewable energy sources. For example, a solar energy source will not be able generate power at nights, which will therefore lower the power availability; a demand response program can offset this effect. This credit rewards projects that move beyond passive energy consumption toward grid-interactive strategies that make buildings smarter, more flexible, and more resilient.

The credit offers four options. Option 1 rewards on-site electric or thermal energy storage with automatic controls that charge during off-peak or low-carbon periods and discharge during high-demand or high-carbon periods. Option 2 requires enrollment in a demand response program, committing the building to reduce its electricity use when called upon by the grid. Option 3 focuses on automated demand-side management, either through system-level controls on major building systems or through a building automation system capable of shedding at least 10% of peak electricity demand in response to grid signals. Option 4 addresses power resilience by requiring the building to island from the grid and sustain critical loads using on-site renewables and energy storage for at least three days.

CREDIT REQUIREMENTS

All projects must evaluate grid-interactive measures relative to the current and forecasted grid context, location, building type, and ownership structure. Interval recording meters and equipment capable of accepting an external signal must be provided. Projects may comply with any combination of Options 1 through 4.

Option 1: Energy Storage (1–2 points)

Provide on-site electric and/or thermal storage with automatic load management controls that store energy during off-peak or low grid carbon intensity periods and discharge during on-peak or high grid carbon intensity periods. Points are based on the ratio of storage capacity to peak demand, with higher capacity earning more points.

AND/OR

Option 2: Demand Response Program (1 point)

Enroll in a minimum one-year demand response contract with a qualified program provider, with the intention of multiyear renewal. On-site electricity generation and fuel combustion cannot be used to meet this requirement.

AND/OR

Option 3: Automated Demand-Side Management (1 point)

Provide automated controls capable of reducing electricity demand in response to grid signals, through one of two paths. Path 1 requires automated demand response controls on at least two major building systems such as HVAC, lighting, receptacles, service water heating, or electric vehicle charging equipment. Path 2 requires a building automation system with a plan to shed at least 10% of peak electricity demand for a minimum of one hour, addressing both winter and summer peaks. On-site electricity generation and fuel combustion cannot be used to meet this requirement.

AND/OR

Option 4: Power Resilience (1 point)

Identify critical equipment requiring continuous operation and design the project to island from the grid, powering critical loads with on-site renewable energy and storage systems for at least three days.

KEY THINGS TO REMEMBER

- Buildings should not be just consumers of energy—they should actively interact with the electrical grid to reduce strain during peak demand, shift loads to cleaner energy periods, and maintain operations when the grid fails.

ENHANCED REFRIGERANT MANAGEMENT—CREDIT

- 1–2 Points (New Construction)

CREDIT SUMMARY

This credit builds on the Fundamental Refrigerant Management prerequisite and rewards project teams that take a step further to limit or eliminate refrigerant impacts. In Option 1, the credit rewards project teams that eliminate refrigerants completely or select low-GWP refrigerants. Option 2, on the other hand, requires project teams to implement stricter measures to reduce refrigerant leakage from refrigerant-containing equipment. Strategies include using self-contained equipment, installing automatic leak detection in enclosed spaces, providing higher-quality piping connections, and creating a refrigerant maintenance plan.

CREDIT REQUIREMENTS

Option 1: No Refrigerants or Low GWP (1–2 points)

Do not use refrigerant-containing equipment.

OR

Demonstrate that the maximum total weighted average refrigerant GWP for all new refrigerant-containing equipment is no greater than 80% of the LEED benchmark values to earn 1 point, and no greater than 50% of the benchmark values to earn 2 points.

Option 2: Limit Refrigerant Leakage (1 point)

Design, construct, and operate the project's refrigerant-using equipment to minimize refrigerant leakage. Requirements include using self-contained equipment (with no field-installed piping) and reducing the opportunities for leaks through better detailing and monitoring. Specify automatic leak detection in fully enclosed spaces. Use brazed or press-type fittings for any field-installed refrigerant piping. Have a refrigerant maintenance plan with an assigned oversight party.

END OF CHAPTER EXERCISES

EXERCISE: MATCHING GAME

Match the following terms to their definition.

Terms		
R-Value	Lighting Power Density	Monitoring-Based Commissioning (MBCx)
Thermal Bridging	Electrification	Retrocommissioning
Thermal Mass	Commissioning	Plug Loads

1) ______________________: The process of using software to monitor real-time data from building automation systems and meters, enabling building operators to detect faulty equipment, identify unusual energy consumption patterns, and address issues as they occur.

2) ______________________: A systematic process for analyzing and optimizing building system performance involving a commissioning provider. It includes testing of building systems and documentation of any measures needed to improve performance.

3) ______________________: Represents the electrical use by all the equipment that is connected to the electrical system via electrical receptacles.

4) ______________________: Indicates how well a material resists heat flow.

5) ______________________: Areas where conductive materials within the building envelope permit unwanted heat transfer in both cold and hot weather conditions.

6) ______________________: The ability of a material to absorb and store energy.

7) ______________________: The amount of power used per square foot or square meter when all lighting is turned on.

8) ______________________: The process of replacing non-electric energy sources with electricity.

9) ______________________: A systematic process carried out by skilled personnel to verify that a building's systems are designed, installed, and operating in accordance with the owner's project requirements (OPR), project goals, and design specifications.

ANSWER: MATCHING GAME

1) Monitoring-Based Commissioning (MBCx)	4) R-Value	7) Lighting Power Density
2) Retrocommissioning	5) Thermal Bridging	8) Electrification
3) Plug Loads	6) Thermal Mass	9) Commissioning

PRACTICE EXAM

To access the end-of-chapter practice exam online in our exam simulator, scan the QR code below with your laptop, phone, tablet, or another compatible device, then log in using your email address and password. If you have trouble scanning the QR code, visit www.projectific.com/account and log in using your email address and password. If you have not yet created a password, please refer to the *How to Access the Exam Simulator* section at the back of the book for instructions.

CHAPTER 9:

MATERIALS AND RESOURCES

Construction and demolition waste constitutes about 40% of the total solid waste stream in the United States.[1] However, there are much better alternatives than classifying items as waste, and there are many ways to reduce the environmental harm associated with materials. Using fewer materials, choosing environmentally preferable materials, using locally harvested materials, and eliminating waste all provide great benefits.

A systems-based, life-cycle perspective and an integrative process will help projects achieve their goals when addressing the use of materials and resources. Life-cycle assessment, which was discussed in the previous chapters, will also serve as a great tool to evaluate materials and resources according to their environmental performances.

This chapter will discuss both the importance of materials and resources for the environment and their effects on the green building design. After discussing the MR credit category, we will go over all the MR prerequisites and credits individually to show how these features are used in LEED certification for a LEED BD+C: New Construction and Major Renovations project.

The MR credit category will be discussed under three major sections:

- Conservation of materials
- Environmentally preferable materials
- Waste management and reduction

CONSERVATION OF MATERIALS

Project teams should approach materials with a mindset of sufficiency, material efficiency, and material circularity. **Sufficiency** refers to measures and practices that avoid unnecessary demand for energy, materials, land, and water while still delivering human well-being within planetary boundaries. **Material efficiency** refers to using less material to perform the same function or achieve the same results. **Material circularity**, on the other hand, refers to a system in which resources are continuously reused and recycled, reducing dependence on virgin materials and minimizing waste.

Rightsizing the project is the first step toward conservation of materials. An example of rightsizing includes designing buildings that are not larger than really needed and will therefore require fewer materials and resources. The same logic also applies to neighborhoods, as denser and more compact mixed-use neighborhoods will require fewer miles of roads and less infrastructure.

Reusing existing materials, salvaged materials, and especially existing buildings results in tremendous material savings. If the project is a major renovation of an existing building, the project team should look for ways to reuse the following existing materials:

- Framing
- Envelopes
- Walls
- Flooring
- Ceilings
- Roofing

Using salvaged materials will contribute to reduction in the demand for virgin materials. Following are some types of salvaged materials that can be used in new buildings:

- Bricks
- Doors
- Windows
- Flooring
- Cabinets, furniture
- Tiles

Adaptive reuse strategies, which can also be called "designing for flexibility," should be well considered in the buildings that may need frequent changes in layouts or floor plans. For example, a flexible floor plan for a hospital building can allow simple changes to be made regarding the sizes of rooms due to the changes in medical equipment that occur alongside the rapid evolution of the technology.

Through the use of movable wall partitions or the use of modular systems, the layout of the hospital plans can be changed without going into a complete renovation.

Using efficient framing techniques can be another strategy to conserve materials by reducing the amount of materials needed. With advanced framing, if studs can be spaced 24 inches apart instead of 16 inches, fewer materials will be needed without reducing structural performance.

Material savings can also come from simplifying interior finishes. In some projects, a polished concrete slab can serve as the finished floor without adding carpet or other flooring materials. Similarly, exposed ductwork or piping can sometimes eliminate the need for additional ceiling materials when the design and building use allow for it. This approach of intentionally reducing the total amount of materials used in a building without compromising its function or performance is known as **dematerialization**.

According to the EPA, the best way to eliminate waste is through **source reduction**. Source reduction refers to the exact sizing of the materials to be produced through prefabrication, modular construction, or similar methods, so that no waste is generated on-site. Since it's about decreasing the unnecessary material brought into a building, it also covers the use of products with less packaging.

Another common source of construction waste is **overage**, which refers to the extra material ordered as a precaution that often goes unused. Minimizing overage through careful design and procurement is ideal, but when surplus materials remain, project teams should look into returning them to the manufacturer or donating them to a reuse outlet rather than sending them to a landfill.

The conservation of materials does not end with the completion of the construction phase. In a green building, the same principles should be applied throughout the building operations phase to the last phase of the building—a demolition or reuse phase.

ENVIRONMENTALLY PREFERABLE MATERIALS

When selecting materials to be used in the project, the project team should consider a material's entire life cycle or life cycles since the aim of a sustainable material should be to have infinite life cycles. After the useful life of a material ends in one building, the material should be used as another product or a part of another product in order to continue being used. In other words, the first goal here is to implement all the strategies to stop the material from going to landfills as waste.

The second goal should be to evaluate the effects of a particular product on the environment. During their life cycles, all buildings and building materials have different impacts on the environment. The building materials are harvested/extracted, manufactured in the factories, and then installed in the buildings to be used. After that, they either get demolished, disposed of, or recycled. Energy is also consumed during each of these stages.

In addition, the way that manufacturers, contractors, or consumers manage each of these steps varies greatly. Materials that are extracted or harvested responsibly, manufactured in environmentally friendly facilities, and can be recycled to become part of another product should be less harmful to the environment than conventional products. Conversely, a product that requires significant energy to extract and lacks durability is unlikely to be considered environmentally friendly.

Using locally produced materials is another plus, as this reduces greenhouse gas emissions that occur during transportation.

The following are some features of environmentally preferable materials:

- Locally harvested/extracted and manufactured
- Sustainably grown and harvested
- Made from **rapidly renewable materials** (which can replenish within 10 years)
- Do not contain any toxins
- Manufactured in factories that support human health and workers' rights
- Long-lasting and reusable
- Contain recycled content
- Have intended end-of-life scenarios that avoid a landfill
- Low in embodied energy

LEED also encourages the use of products with third-party certifications that verify responsible sourcing and sustainable practices. The **Forest Stewardship Council** (**FSC**) certification is one such example, widely recognized for verifying that wood and wood-based materials come from forests managed in an environmentally responsible, socially beneficial, and economically viable manner. FSC provides a chain-of-custody certification, which tracks a product from the forest through every stage of production and distribution, ensuring that the certified material is properly identified throughout the supply chain.

To take the discussion to the next level, the following topics will describe the tools and processes that will enable the project teams to make the best decisions on selecting environmentally preferable materials.

LIFE-CYCLE ASSESSMENT

To examine the environmental cost of a building material, we need to consider the carbon emissions resulting from the material's manufacturing, transportation, installation, and disposal—in other words, its embodied carbon. By way of illustration, consider marble that is mined in China, manufactured in Europe, transported to the United States for purchase, and then used in a building. The embodied carbon of that marble includes all the carbon emissions released during extraction, manufacturing, transportation, installation at the site, and, finally, disposal. The more embodied carbon a material has, the more it contributes to environmental damage.

In addition to carbon emissions, building materials create other forms of pollution. As factories refine raw materials into usable building products, they can release pollutants into the air and water, affecting workers, nearby communities, rivers, and habitats along the entire supply chain. Even after the materials arrive at the project site, the construction process adds its own layer of impact. A construction site runs on energy—heavy equipment, delivery vehicles, heaters, lights, power tools, and more. When grid electricity is not available, generators are brought in to power the work, and those generators often run on fossil fuels unless battery-powered electrical equipment is used. But how can project teams systematically compare one material against another across all of these environmental impacts? This is where life-cycle assessment comes in.

Life-cycle assessment (**LCA**), which quantitatively examines all the environmental effects of a product—or even a building—over its entire life cycle, allows project teams to understand the background of building products. Project teams can evaluate the trade-offs between different materials and select those that are the best fit for both the project and the environment. In LCA, a cradle-to-grave approach, and even more comprehensively, a cradle-to-cradle approach, may be used, allowing both the total energy use and other environmental consequences resulting from the creation of a material to be calculated.

A **cradle-to-cradle assessment** analyzes a product's life cycle from the initial extraction or harvesting of its resources to its new life as another product or as part of another product. Thus, cradle-to-cradle products are waste-free products that can be recycled—because of their high recyclable content—and reused.

By contrast, a **cradle-to-grave assessment** analyzes a product's life cycle from the initial extraction or harvesting of its resources to the end of its life, resulting in waste. Since cradle-to-grave products cannot be recycled, they have only one life cycle and ultimately end up in landfills.

For exam purposes, it is also important to be aware of **cradle-to-gate assessment**, which some manufacturers choose to conduct in order to evaluate a product's partial life cycle—from resource extraction or harvesting to its manufacture as a product ready for sale at the factory gate.

Life Cycle Stages															
Product Stage			Construction Stage		Use Stage					End of Life Stage				Beyond
A1	A2	A3	A4	A5	B1	B2	B3	B4	B5	C1	C2	C3	C4	D
Raw Materials Supply	Transport	Manufacturing	Transport	Construction & Installation	Use	Maintenance	Repair	Replacement	Refurbishment	Deconstruction / Demolition	Transport	Waste Processing	Disposal	Reuse / Recovery / Recycling Potential
					B6 Operational Energy									
					B7 Operational Water									

← Cradle to Gate →

← Cradle to Grave →

← Cradle to Cradle →

The **Cradle to Cradle Products Innovation Institute** developed a Cradle to Cradle certification that evaluates products across five sustainability categories: material health, product circularity, clean air and climate protection, water and soil stewardship, and social fairness. Products are rated at four levels — Bronze, Silver, Gold, or Platinum — with the overall rating determined by the lowest level achieved across the five categories. Cradle to Cradle certified products have demonstrated measurable achievements in sustainability.

MATERIAL INGREDIENTS AND THEIR EFFECTS ON HUMAN HEALTH

Building materials can contain substances that are harmful to human health. These hazards can affect people at every stage of a product's life cycle—from the workers who extract raw materials and manufacture products, to the construction crews who install them, to the occupants who live and work around them for years.

As will also be discussed under the Low-Emitting Materials credit, one of the biggest indoor air quality concerns is volatile organic compounds. **Volatile organic compounds** (**VOCs**) are chemicals that evaporate at room temperature and can be released from a wide range of building products, including paints, adhesives, sealants, flooring, and furniture. While some VOCs are relatively harmless, others can cause headaches in the short term and serious conditions such as cancer over time. Composite wood products like plywood and particleboard are a particular concern because they are commonly manufactured with formaldehyde-based adhesives, making **formaldehyde** one of the most critical indoor pollutants to control.

Beyond VOCs, many building materials contain other hazardous substances that deserve attention. Heavy metals such as **lead** and **chromium** are found in plumbing, wiring, and roofing products. **PFAS**, often called "**forever chemicals**" because they do not break down in the environment, show up in certain paints and stain-resistant fabric coatings.

Not every hazardous substance in a building product ends up in the finished material—some are only used during the manufacturing process. Even so, these chemicals can still harm workers in factories, contaminate nearby water sources, and affect communities along the supply chain. This is why understanding what goes into our building materials and choosing healthier alternatives is a critical responsibility for every project team.

PRODUCT TRANSPARENCY

Project teams should be careful about **greenwashing** when selecting building products. Greenwashing refers to presenting a product as more environmentally friendly than it actually is. To avoid this, environmentally preferable materials should provide accurate and transparent information and declarations to buyers.

One example is the **environmental product declaration** (**EPD**), which reports a product's environmental impacts based on life-cycle assessment data. An EPD describes the environmental effects of raw material extraction, energy use, chemical composition, waste generation, and emissions to water, air, and soil throughout the product's sourcing and manufacturing process.

Similarly, a product with a **health product declaration** (**HPD**) provides disclosure about its material ingredients, potential chemical hazards, related concerns, and additional health information.

By considering products with transparent information and declarations, project teams can make better decisions.

WASTE MANAGEMENT AND REDUCTION

The construction industry generates huge amounts of waste. Green building projects should work on reducing the amount that ends up in landfills through the development and implementation of a **waste management plan** for both the construction and operation phases of a building.

If an older building is already on the site, the smartest move during construction is to treat it like a material "warehouse," not a pile of debris. Deconstruction—taking the building apart—can salvage far more usable material than demolition, and the same logic applies during renovations. Whatever cannot be reused should be sorted on-site so it can be recycled. Anything not separated on-site typically goes into a mixed-waste bin, where it can still be recycled if a local processing facility is available.

During building operations, waste reduction starts long before the building opens—during building design. Project teams should think about which waste streams will be present and in what quantities, what local hauling and processing services exist, and how much effort each waste stream requires to manage. All collection points and pathways should be designed to make sorting easy for occupants and staff and pickup convenient for waste haulers. After occupancy, facility managers should track waste by source and material type, then reduce it by improving internal reuse practices, setting up composting where possible, and maximizing recycling and diversion. The overall goal, in both construction and operations, is to keep waste from ending up in landfills or waste-to-energy systems.

STRATEGIES TO ADDRESS MATERIALS AND RESOURCES

- Approach materials with a mindset of sufficiency, material efficiency, and material circularity.
- Right-size the project.
- Reuse existing buildings and salvaged materials.
- Recycle products and materials that cannot be reused.
- Select durable, low-maintenance, locally sourced materials with low embodied carbon.
- Use adaptive reuse strategies for buildings that may need frequent changes in layouts or floor plans.
- Implement source reduction and exact sizing through prefabrication, modular construction, or similar methods so that less waste is generated on-site.
- Minimize overage through careful design and procurement. If surplus materials remain, return them to the manufacturer or donate them to a reuse outlet instead of sending them to a landfill.
- Conduct life-cycle assessment (LCA) to quantitatively examine the environmental effects of building products over their entire life cycle and better understand their impacts.
- Look for Cradle-to-Cradle certified products.
- Select products that are sustainably grown and harvested, made from rapidly renewable materials, toxin free, manufactured in factories that support human health and workers' rights, long-lasting and reusable, contain recycled content, have end-of-life scenarios that avoid landfills, and have low embodied energy.
- Verify material ingredients and consider their effects on human health.
- Pay attention to VOCs and select low-emitting and non-emitting products when possible.
- Pay attention to heavy metals such as lead and chromium in plumbing, wiring, and roofing products.
- Avoid greenwashing and look for transparent products that include an EPD and an HPD.
- Implement a waste management plan for both the construction and building operation phases.
- If grid electricity is not available, switch from diesel-powered equipment to battery-powered equipment.

MATERIALS AND RESOURCES PREREQUISITES AND CREDITS

PLANNING FOR ZERO WASTE OPERATIONS–PREREQUISITE

PREREQUISITE SUMMARY

Waste is one of the primary burdens on the environment. In the United States, metals, paper, food, glass, and plastics comprise approximately 69% of the total amount of municipal solid waste.[2] Furthermore, all the mentioned products can be recycled, and recycling not only reduces the need for virgin materials, but it also reduces the need for landfills.

In addition to solid waste, electronic waste is a major problem in today's world, since the disposal of these materials can cause more harm to the environment. But handling electronic waste requires different procedures.

This prerequisite aims to reduce the waste that is generated by building occupants. (Waste produced during the construction phase will be addressed by the Construction and Demolition Waste Diversion credit.) In turn, this prerequisite forces project teams to have spaces dedicated to the collection and storage of recyclables, including organics/food waste, mixed paper, corrugated cardboard, glass, plastics, and metals.

Projects should also take appropriate measures for the safe collection, storage, and disposal of batteries, mercury-containing lamps, and electronic waste.

PREREQUISITE REQUIREMENTS

Project teams must meet both of the following requirements.

1) Storage and Collection of Recyclables:

Provide dedicated collection and storage areas for recyclable materials for the entire building, and make sure these areas are accessible to waste haulers, janitorial staff, and building occupants. Collection and storage may be in separate locations.

The recycling program must include space for organics/food waste, mixed paper, corrugated cardboard, glass, plastics, and metals. Mixed recyclables are acceptable for paper, cardboard, glass, plastics, and metals where local collection systems require it.

Also take appropriate measures for the safe collection, storage, and disposal of batteries, mercury-containing lamps, and electronic waste.

AND

2) Zero Waste Operations Planning:

Provide design details, maintenance manuals, and/or other resources from the design and construction team that help occupants and operators achieve high-performance waste prevention and recycling once the building is in operation.

Core and Shell projects should communicate the building's infrastructure and service options in the tenant guidelines.

KEY THINGS TO REMEMBER

- This prerequisite addresses waste generated during the building operations phase, not the construction phase. The Construction and Demolition Waste Diversion credit, on the other hand, addresses waste generated during the construction phase.
- With this prerequisite, every LEED project must have a recycling program. The recycling program must cover organics/food waste, mixed paper, corrugated cardboard, glass, plastics, and metals.
- Every LEED project should take appropriate measures for the safe collection, storage, and disposal of batteries, mercury-containing lamps, and electronic waste.
- Core and Shell projects should communicate the building's waste management system in the tenant guidelines.

QUANTIFY AND ASSESS EMBODIED CARBON—PREREQUISITE

PREREQUISITE SUMMARY

As building material decisions are most flexible early in a project, teams can make the biggest meaningful reductions in embodied carbon during schematic design and design development. A collaborative process among the owner, designers, and contractors helps teams choose building materials with reduced environmental impact.

Note that, as its name implies, the goal of this prerequisite is to quantify and assess the embodied carbon of building materials to increase awareness of the impacts of material choices. It does not mandate reductions in embodied carbon. The prerequisite focuses on quantifying cradle-to-gate (A1–A3) embodied carbon emissions for each building material, which include emissions associated with raw material extraction (A1), transport to the manufacturer (A2), and manufacturing (A3). To quantify the embodied carbon, teams will multiply the global warming potential (GWP) of each building material (from Environmental Product Declarations or default values) by the quantity of each material used.

After completing the calculations, projects must identify the three biggest contributors to embodied carbon. For each key source, teams should describe the strategies they evaluated and applied to reduce impacts—for instance, choosing lower-carbon concrete mixes, selecting materials with higher recycled content, or adjusting the design to use less material overall. By focusing on these hot spots, teams can target changes that have the greatest potential to reduce embodied carbon.

PREREQUISITE REQUIREMENTS

Project teams must meet both of the following requirements.

1) Embodied Carbon:

Quantify the embodied carbon impacts—global warming potential (GWP)—of the project's structure, enclosure, and hardscape materials. Include all ancillary structures within the LEED project boundary, such as parking structures or outbuildings.

At a minimum, include the following material categories in the calculations: asphalt, concrete, masonry, structural steel, insulation, aluminum extrusions, structural wood and composites, cladding, and glass.

Quantify cradle-to-gate (A1–A3) embodied carbon emissions for each material, calculated as the product's GWP per unit multiplied by the quantity used.

AND

2) High-Priority Embodied Carbon Sources:

Identify the top three sources of embodied carbon on the project and describe how project-specific strategies were considered to reduce the impacts of those sources.

Core and Shell projects should include embodied carbon measurements and material suppliers in the tenant guidelines.

KEY THINGS TO REMEMBER

- LEED requires all projects to quantify and assess the embodied carbon of building materials to increase awareness of the impacts of material choices. To do so, it requires quantifying cradle-to-gate (A1–A3) embodied carbon emissions for each building material, which include emissions associated with raw material extraction (A1), transport to the manufacturer (A2), and manufacturing (A3).
- Because building material decisions are most flexible early in a project, teams can make the most meaningful reductions in embodied carbon during schematic design and design development.

BUILDING AND MATERIALS REUSE—CREDIT

- 1–3 Points (New Construction)

CREDIT SUMMARY

Producing new building materials requires raw resource extraction, energy, transportation, and processing—and all of that adds up to embodied carbon and pollution. Nevertheless, there is a more effective way to reduce the environmental impact of a building, which is to reuse existing buildings and materials, since it is not necessary to consider the environmental impacts of producing new materials when existing ones can be reused.

Option 1 of this credit rewards projects that keep the existing building structure, including floor decking, roof decking, and enclosure, and points are awarded based on how much of the existing structure and enclosure is reused by project area. Option 2, on the other hand, rewards projects that identify and reuse materials either by keeping them in place or by bringing in reused materials from salvage sources and reuse markets.

CREDIT REQUIREMENTS

Option 1: Building Reuse (1–3 points)

Maintain the existing building structure, including floor decking, roof decking, and enclosure. Points are awarded based on the percentage of existing structure and enclosure reused by project area. Any portions of the existing building that are deemed structurally unsound or hazardous are excluded from the credit calculations.

AND/OR

Option 2: Materials Reuse (1–2 points)

Survey the project to identify opportunities to reuse materials in place and/or to procure reused materials from off-site. Reused materials can come from on-site retention, salvage sources, or reuse markets, and they must be incorporated into the new project design.

If deconstruction or demolition is part of the scope, complete a salvage assessment before work begins to identify materials that can be kept on-site or diverted off-site to reuse markets. Salvaged materials sent off-site for reuse contribute to the Construction and Demolition Waste Diversion credit, while materials retained on-site contribute to this option. Points are awarded based on the reuse percentage for each material type.

REDUCE EMBODIED CARBON—CREDIT

- 1–6 Points (New Construction)

CREDIT SUMMARY

This credit rewards projects for tracking and reducing embodied carbon associated with major structural, enclosure, and hardscape materials.

In Option 1, projects may earn points by completing a cradle-to-grave whole-building life-cycle assessment for the building's structure, enclosure, and hardscape materials, and comparing the results to a baseline design. As discussed, a cradle-to-grave framework investigates the environmental impacts of products or materials from their extraction to their disposal. A **cradle-to-grave whole-building life-cycle assessment** (**WBLCA**), on the other hand, quantifies the environmental impacts of a building during its life, from its construction to occupancy to its eventual demolition.

In Option 2, project teams can earn points through environmental product declaration analysis. Points may be achieved by reducing the embodied carbon of the project using environmental product declaration (EPD) data for procured materials compared with industry-average values. As mentioned, environmental product declarations (EPDs) report a product's environmental impacts based on life-cycle assessment data. An EPD shows the product's raw material extraction effects on the environment, energy use, chemical makeup, waste generation, and any emissions made to water, air, and soil during the entire sourcing and manufacturing process.

Alternatively, in option 2, project teams can earn points by demonstrating that the targeted structural, enclosure, and hardscape material types have lower embodied carbon impacts than industry benchmarks using product-specific Type III EPDs. A **Type III EPD** is a third-party verified EPD prepared in accordance with international standards and product category rules. It is considered the most rigorous form of EPD because it provides standardized, independently verified environmental data.

At last, in Option 3 of this credit, project teams can earn points by tracking carbon emissions from construction activities based on documented fuel and utility use. Points are awarded depending on whether tracking covers contractor jobsite operations only or both contractor and subcontractor jobsite operations.

CREDIT REQUIREMENTS

Quantify the reduction of embodied carbon of major structure, enclosure, and hardscape materials. Any ancillary structures within the LEED project boundary, including parking structures or outbuildings, must be included in the credit calculations.

Option 1: Whole-Building Life-Cycle Assessment (1–6 points)

Conduct a cradle-to-grave whole-building life-cycle assessment (WBLCA) for the project's structure, enclosure, and hardscape materials. Compare the results to a project baseline and earn points based on the percent reduction.

The WBLCA report must include results for these impact categories:

- Global warming potential (GWP)
- Depletion of the stratospheric ozone layer
- Acidification of land and water sources
- Eutrophication
- Formation of tropospheric ozone
- Depletion of nonrenewable energy resources

AND/OR

Option 2: Environmental Product Declaration (EPD) Analysis (1–3 points)

Under this option, project teams have two paths. In Path 1, project teams can earn points for reducing the embodied carbon of the project based on EPD data for procured materials compared to industry average values. Alternatively, in Path 2, project teams can earn points by demonstrating that structural, enclosure, and hardscape materials for targeted material types have lower embodied carbon impacts than industry benchmarks, as demonstrated by product-specific Type III EPDs.

AND/OR

Option 3: Track Carbon Emissions from Construction Activities (1–2 points)

Earn points by tracking carbon emissions from construction activities based on documented fuel and utility use. Points are awarded depending on whether tracking covers contractor jobsite operations only or both contractor and subcontractor jobsite operations.

KEY THINGS TO REMEMBER

- Cradle-to-grave framework investigates the environmental impacts of products or materials from their extraction to their disposal, and cradle-to-grave WBLCA quantifies the environmental impacts of a building during its life, from its construction to occupancy to its eventual demolition.
- Type III EPDs are third-party verified, prepared according to international standards and product category rules, and are considered the most rigorous form of EPD.

LOW-EMITTING MATERIALS—CREDIT

- 1–2 Points (New Construction)

CREDIT SUMMARY

The levels of volatile organic compounds (VOCs) in building materials are important to address because, during the installation and operation phases, these materials release VOCs indoors. Consequently, they can pose a serious health hazard to both construction workers and building occupants.

Think about epoxy-based paints. During the painting process, their smell is not pleasant at all. They release very strong odors, which can even create a burning feeling in the throat and lungs. This is caused by a high VOC level. Materials with a high level of VOCs can cause asthma, chronic obstructive pulmonary disease, and cancer, while also causing harm to the environment.

The credit requires certain VOC levels to be met for paints and coatings, flooring, ceilings, adhesives and sealants, walls, insulation, composite wood, and furniture. Additionally, the credit sets criteria for evaluating formaldehyde emissions.

Inherently nonemitting sources, which are products without any VOC emissions, are exempt from this credit's requirements. Examples of inherently nonemitting sources are ceramic, stone, powder-coated metals, concrete, glass, anodized metal, clay brick, and untreated solid wood.

CREDIT REQUIREMENTS

Specify and install permanently installed products, paints, coatings, adhesives, sealants, flooring, walls, ceilings, insulation, furniture, and/or composite wood products that meet the following criteria.

1) VOC emissions evaluation criteria:

A product must have a qualifying third-party certification demonstrating compliance with California Department of Public Health (CDPH) Standard Method v1.2-2017. Alternatively, the product may have a qualifying independent laboratory report. Products that are inherently non-emitting, salvaged, or reused are considered compliant.

2) Furniture emissions evaluation criteria:

Furniture must hold a qualifying third-party certification demonstrating testing per ANSI/BIFMA Standard Method M7.1-2011. Products that are inherently non-emitting, salvaged, or reused are considered compliant.

3) Formaldehyde emissions evaluation criteria:

Composite wood products must have a qualifying third-party certification from a California Air Resources Board (CARB) or EPA-recognized third-party certifier. Products that are inherently non-emitting, salvaged, or reused are considered compliant.

KEY THINGS TO REMEMBER

- Inherently nonemitting sources, which are products that do not emit VOCs, are exempt from LEED's low-emitting materials requirements. Examples of inherently nonemitting sources are ceramic, stone, powder-coated metals, concrete, glass, anodized metal, clay brick, and untreated solid wood.

BUILDING PRODUCT SELECTION AND PROCUREMENT—CREDIT

- 1–5 Points (New Construction)

CREDIT SUMMARY

Every building is assembled from thousands of individual products and each of these products carries an environmental footprint, from raw material extraction and manufacturing emissions to worker health conditions and end-of-life disposal.

This credit recognizes that the choices project teams make at the product selection stage have a profound ripple effect across the entire supply chain.

This credit rewards teams for selecting nonstructural building products that demonstrate verified sustainability attributes across five criteria areas: climate health, human health, ecosystem health, social health and equity, and circular economy.

Rather than focusing on a single environmental metric, the credit takes a multi-attribute approach—encouraging manufacturers to disclose sustainability information and optimize their products across multiple dimensions simultaneously.

CREDIT REQUIREMENTS

Project teams should select nonstructural building products that demonstrate sustainability achievement in one or more of the following five criteria areas:

- Climate health
- Human health
- Ecosystem health
- Social health and equity
- Circular economy

A product that achieves two or more criteria areas is considered multi-attribute.

Achievement must be shown through eligible manufacturer product documentation, such as third-party certifications, ecolabels, declarations, and standards.

Products are scored at three achievement levels.

Level 1 products earn a 1× multiplier and represent a first step toward sustainability within a criteria area. Level 2 products earn a 2× multiplier and represent a leadership position in the marketplace. Level 3 products earn a 3× multiplier and represent elite, forefront sustainability performance. A product's scores across all applicable criteria areas are summed to produce its multi-attribute score (MAS), capped at a maximum of 5.

Each product's value—measured by cost, area, volume, or unit—is multiplied by its MAS to determine its adjusted product value for LEED. Within each product category, the adjusted values of all compliant products are summed and divided by the total value of all products in that category. When the resulting percentage equals or exceeds 100%, the project earns 1 point for that category.

Points are awarded for whole product categories, up to a maximum of 5 points for 5 qualifying categories.

CONSTRUCTION AND DEMOLITION WASTE DIVERSION—CREDIT

- 1–2 Points (New Construction)

CREDIT SUMMARY

Every ton of construction and demolition materials kept out of a landfill or incineration represents both a reduction in environmental harm and an opportunity to recapture material value through recycling, salvage, and reuse. The goal of this credit is to develop and implement strategies to prevent construction and demolition materials from becoming waste.

The credit requires project teams to develop a construction and demolition (C&D) materials management plan early in the design phase. This plan sets out the strategies, target diversion rates, and logistics for how waste will be handled on site. Second, the team must execute that plan and document results in a final waste management report showing all waste generated and the project's disposal and diversion rates. Points are awarded based on the percentage of the total construction and demolition material diverted from landfill and incineration.

Acceptable diversion strategies include off-site salvage, source separation for single-material recycling, mixed recycling, and manufacturer or industry take-back programs. Note that as source-separated materials are not mixed with other materials, they have higher overall diversion rates. As a result, LEED encourages project teams to set up dedicated areas for each construction and demolition material.

Certain materials, such as alternative daily cover, are excluded from credit calculations. Alternative daily cover (ADC) refers to materials used to cover the active face of a landfill at the end of each operating day as a substitute for soil. Because the material ends up permanently embedded in a landfill, LEED classifies alternative daily cover as disposal rather than a diversion strategy.

Common construction and demolition waste materials to be diverted include concrete, metals, wood, brick, cardboard, gypsum board, ceilings, furniture, and carpet.

CREDIT REQUIREMENTS

Create and implement a construction and demolition (C&D) materials management plan and earn points through diversion and recycling. Submit a final waste management report showing all waste generated and the project's disposal and diversion rates. Points are awarded based on the percentage of the total construction and demolition material diverted.

To divert construction and demolition waste, project teams may use strategies such as off-site salvage, source separation for single-material recycling, mixed recycling, and manufacturer/industry take-back programs.

Source-separated materials are counted as 100% diverted. This includes recovered materials sent to a single-material recycler, recovered materials sent for off-site salvage/reuse, and materials sent to approved take-back programs. On the other hand, mixed construction and demolition materials sent to a processing facility for recovery must take the facility's average recycling rate as such materials have lower recycling rates compared to source-separated materials.

Materials sent for alternative daily cover or incineration/energy recovery are treated as waste (0% diverted).

Hazardous waste, excavated soil, and land-clearing debris are also excluded from calculations.

KEY THINGS TO REMEMBER

- LEED rewards projects that create and implement a construction and demolition materials management plan early in the project and awards points based on the percentage of the total construction and demolition material diverted from landfill and incineration.
- To divert construction and demolition waste, LEED encourages project teams to use strategies such as off-site salvage, source separation for single-material recycling, mixed recycling, and manufacturer/industry take-back programs.
- Hazardous waste, excavated soil, and land-clearing debris are excluded from LEED waste diversion calculations. Additionally, materials sent for alternative daily cover or incineration/energy recovery are treated as waste. Common construction and demolition waste materials to be diverted include concrete, metals, wood, brick, cardboard, gypsum board, ceilings, furniture, and carpet.

END OF CHAPTER EXERCISES

EXERCISE: MATCHING GAME

Match the following terms to their definition.

Terms		
Sufficiency	Source Reduction	Environmental Product Declaration (EPD)
Material Efficiency	Overage	Volatile Organic Compounds (VOCs)
Material Circularity	Cradle-To-Cradle	

1) ________________________: The exact sizing of materials through prefabrication, modular construction, or similar methods to prevent waste generation on-site. It also covers the use of products with less packaging.

2) ________________________: Extra material ordered as a precaution during construction that often goes unused.

3) ________________________: The practice of designing materials to have "infinite" life cycles by recovering and recycling them into new products, creating a closed-loop system instead of waste.

4) ________________________: A set of measures and practices that avoid unnecessary demand for energy, materials, land, and water while still delivering human well-being within planetary boundaries.

5) ________________________: Using less material to perform the same function or achieve the same results.

6) ________________________: A system in which resources are continuously reused and recycled, reducing dependence on virgin materials and minimizing waste.

7) ________________________: Chemicals that evaporate at room temperature and can be released from building products such as paints, adhesives, sealants, flooring, and furniture.

8) ________________________: Disclosure that reports a product's environmental impacts based on life-cycle assessment data.

ANSWER: MATCHING GAME

1) Source Reduction	4) Sufficiency	7) Volatile Organic Compounds (VOCs)
2) Overage	5) Material Efficiency	8) Environmental Product Declaration (EPD)
3) Cradle-To-Cradle Assessment	6) Material Circularity	

PRACTICE EXAM

To access the end-of-chapter practice exam online in our exam simulator, scan the QR code below with your laptop, phone, tablet, or another compatible device, then log in using your email address and password. If you have trouble scanning the QR code, visit www.projectific.com/account and log in using your email address and password. If you have not yet created a password, please refer to the *How to Access the Exam Simulator* section at the back of the book for instructions.

CHAPTER 10: INDOOR ENVIRONMENTAL QUALITY

This credit category is about the strategies related to indoor air quality and thermal, visual, and acoustic comfort. A good standard of indoor environmental quality is essential in green buildings to protect the health and comfort of building occupants. Moreover, thermal, visual, and acoustic qualities of a building are important in increasing occupants' comfort level and productivity.

This chapter will discuss both the importance of indoor environmental quality for the building occupants and construction workers and its effects on the green building design. After discussing the EQ credit category, we will go over all the EQ prerequisites and credits individually to show how these features are used in LEED certification for a LEED BD+C: New Construction and Major Renovations project.

The EQ credit category will be discussed under two major sections:

- Indoor air quality
- Increased occupant comfort

INDOOR AIR QUALITY

In order to increase the indoor air quality of a building, first, the sources of contaminants should be identified, and necessary measures should be taken to prevent them from reaching the indoors, which is also called **source control**.

Following are some of the contaminants that will negatively affect the indoor air quality of a building:

- Tobacco smoke
- Building materials with high levels of volatile organic compounds (VOCs)
- Mold growth
- Cleaning products
- Construction contamination of building materials
- Combustion process in the HVAC equipment
- Radon or methane off-gassing from the building's located soil
- Pollutants in the occupants' shoes
- Pollutants from specific processes used in hospitals, laboratories, and factories

Studies show that poor indoor air quality can result in respiratory disease, allergies, asthma, and sick building syndrome (SBS) or other building-related illnesses (BRI).

In order to avoid poor indoor air quality, a project should do the following:

- Set minimum air quality standards
- Stop secondhand smoke
- Use non-emitting or low-emitting materials
- Protect the site during construction
- Schedule construction activities to minimize occupant exposure
- Increase ventilation rates
- Monitor carbon dioxide
- Implement green cleaning practices
- Periodically maintain and replace air filters
- Reduce radon levels

To take the discussion to the next level, let's discuss these strategies individually.

SET MINIMUM AIR QUALITY STANDARDS

All LEED BD+C projects must comply with certain aspects of the ASHRAE 62.1-2022 standard (a ventilation standard used to achieve acceptable indoor air quality) or the local codes as part of the Fundamental Air Quality prerequisite.

After the minimum air quality standards have been set, monitoring indoor air quality becomes essential because, in some cases, simply bringing "enough" outside air into the building may still not deliver healthy air indoors. For example, if a project is located in a city with high outdoor air pollution, the filtration level of the supply air may need to be increased, and additional testing or verification may be necessary to confirm that the building is maintaining high indoor air quality.

STOP SECONDHAND SMOKE

Exposure to **environmental tobacco smoke** (**ETS**), which refers to secondhand smoke from another person's cigarette, cigar, pipe, or electronic smoking device, can cause lung cancer in nonsmokers and increase the risk of heart disease.

Any school project that aims for a LEED certification should prohibit smoking on the entire campus, both indoors and outdoors.

For other project types, LEED prohibits smoking indoors and does not allow smoking within 25 feet of the building entrance.

For residential projects, LEED requires partitioning of the rooms and also prohibits smoking in the common spaces. Additionally, it requires a blower door test to ensure that air cannot leak from smoking units to others.

USE NON-EMITTING OR LOW-EMITTING MATERIALS

Material off-gassing can seriously affect indoor air quality and human health. The amount of volatile organic compounds (VOCs), such as formaldehyde, contained in construction materials will affect the health of both the construction workers and the building occupants.

Many building materials, such as adhesives, sealants, composite wood, paints, coatings, floorings, furniture, wall coverings, and other products such as photocopy machines can contain harmful levels of VOCs.

LEED addresses VOC and other requirements regarding material off-gassing under the Low-Emitting Materials credit, and in order to earn the credit, projects need to follow the VOC thresholds set for each item category.

Naturally occurring materials and products made from inorganic materials such as marble or granite contain low or no VOCs and are therefore exempt from LEED requirements. These materials are called **inherently nonemitting materials**.

PROTECT THE SITE DURING CONSTRUCTION

The project team should develop and implement an **indoor air quality** (**IAQ**) **plan** in order to protect the health of both the construction workers and the building occupants. While developing the IAQ plan, the project team can refer to the **Sheet Metal and Air Conditioning National Contractors Association (SMACNA)** guidelines, which describe the necessary control measures to be taken during the construction phase to maintain sufficient indoor air quality.

Think about the HVAC ducts stored on-site without any protection and collecting all the dust before getting installed; even if the building has increased its ventilation rates, all the air that will be supplied by the ducts will be contaminated air, which will decrease the indoor air quality as a consequence.

SMACNA recommends practical construction-phase controls to protect indoor air quality, including protecting HVAC equipment and ductwork from dust and moisture, limiting pollutant sources, and using barriers to keep contaminants from spreading into clean or occupied areas. It also emphasizes good housekeeping and careful scheduling so dirty activities happen at the right time and are cleaned up properly.

To ensure compliance, LEED requires project teams to take and send site photos to document their actions.

SCHEDULE CONSTRUCTION ACTIVITIES TO MINIMIZE OCCUPANT EXPOSURE

After the completion of construction, the project team should make sure that any contamination and dust are cleared before the building occupants arrive. To ensure sufficient air quality after the completion of construction, an **air quality test** can be conducted.

Or the project team can choose to conduct a **flush-out**, which is a process of supplying ample amounts of fresh air to the building before or during occupancy; this process will take away the contaminated air and establish the desired level of indoor air quality.

INCREASE VENTILATION RATES

Fresh air is by far the most important element of indoor air quality, and supplying ample amounts of fresh air inside the building will strongly affect the indoor air quality. This can be established by increasing the ventilation rates of the HVAC system. Increased ventilation will result in lowering the ratio of poor air and increasing the removal of indoor contaminants.

However, pulling outdoor air into indoor areas requires fan energy, and that outdoor air typically needs conditioning—heating, cooling, humidity control, and filtration. As this conditioning can be energy-intensive, the goal is to provide the right amount of fresh air for health and comfort—without overventilating and wasting energy. For that reason, most HVAC systems recirculate a portion of conditioned indoor air multiple times, using filters to capture contaminants picked up indoors, while also introducing a controlled amount of fresh outdoor air to dilute pollutants. The recirculation percentage depends on many factors, including the building type and space function. In some healthcare spaces, for example, codes and standards may require 100% exhaust or no recirculation to prevent cross-contamination. Many other building types, on the other hand, commonly recirculate a significant portion of indoor air and add a smaller percentage of fresh outdoor air to meet ventilation requirements.

Projects can further reduce energy use by adjusting ventilation rates to match typical occupancy patterns. For instance, a building may be designed with different air-exchange rates during occupied and unoccupied hours. **Demand-controlled ventilation** takes this idea one step further by automatically adjusting outdoor airflow based on real-time signals from sensors. Most systems use carbon dioxide as a practical indicator of occupancy: when CO_2 levels rise, the system increases ventilation; when levels drop, it reduces ventilation. The goal is to provide the right amount of fresh air for health and comfort—without overventilating and wasting energy.

If the climate in the project location is appropriate, **natural ventilation** can be established by the opening of windows and doors, which would also reduce the building operating costs.

In an appropriate climate, projects can use **mixed-mode ventilation,** which employs both HVAC and natural ventilation.

MONITOR CARBON DIOXIDE (CO_2)

An effective way to ensure good indoor air quality is to monitor the carbon dioxide (CO_2) levels inside the building. By installing carbon dioxide monitoring equipment, the thresholds for CO_2 levels can be determined, and once CO_2 levels increase above the threshold, necessary actions can be taken to increase the amount of fresh air inside the building.

IMPLEMENT GREEN CLEANING PRACTICES

Using green cleaning products will have a lower detrimental impact than regular cleaning products. A **green cleaning program** can be developed to specify:

- the green cleaning products to be used, such as those meeting Green Seal, Environmental Choice, or EPA standards;
- the chemicals permitted for use inside the building;
- training requirements for cleaning personnel on the use of chemicals and green cleaning practices;
- indoor pest control plans; and
- energy-efficient cleaning equipment.

Additionally, installing cleanable entryway grates, grilles, or mats at every building entrance can keep the dirt, water, and contamination from shoes away from the building. These strategies will increase the indoor air quality while also reducing the amount of cleaning chemicals used in the building.

PERIODICALLY MAINTAIN AND REPLACE AIR FILTERS

Without quality air filtration media, no building can establish good indoor air quality. The **minimum efficiency reporting value (MERV)** rates the air filters according to their success in removing particles from the air. Air filters with higher MERV values will remove more particles from the air, but more energy will be consumed in passing the air through them.

To avoid clogged filters, frequent maintenance and replacement of air filters is necessary. Clogged filters will slow the velocity of air inside the ducts, cause molds, and also increase energy consumption of the HVAC equipment.

REDUCE RADON LEVELS

Radon is a radioactive gas that is naturally found in soils, rocks, and water bodies. Although we cannot see, smell, or taste radon, it can be detected with special equipment. If radon is found in a building, it is a serious problem that requires urgent action.

The presence of radon inside a building can create serious health effects such as lung cancer.

Radon can infiltrate a building through the following:

- Cracks in solid floors
- Cracks in walls
- Construction joints
- Gaps in suspended floors
- Gaps around service pipes
- Water supply
- Cavities inside walls

INCREASE OCCUPANT COMFORT

Building occupants need to feel comfortable and in control of their environment in order to be healthy, happy, and productive in their buildings. There are several factors that will affect the comfort of an occupant. These include the following:

- Thermal comfort
- Daylighting
- Views
- Acoustics
- Lighting controls
- Temperature controls
- Ergonomics

However, merely addressing these factors will not be enough. **Occupant surveys** are also essential to ensure occupant comfort. By receiving occupant feedback, the building management can pinpoint areas that need to be improved.

THERMAL COMFORT

Thermal comfort involves much more than simply adjusting the room temperature. It also includes air movement and humidity. An occupant cannot feel comfortable under an HVAC duct that blows directly on them or in a space where the system does not provide even temperatures throughout.

DAYLIGHTING

Daylighting, a very important resource to use in green buildings, will increase the occupants' health, comfort, and productivity while also reducing building operating costs by using less artificial lighting inside the building. While too much daylighting can create discomfort, through means such as glare or direct sunlight to the occupant, an effective lighting design can overcome these consequences.

Increased cooling loads resulting from the heat of daylighting may be another consequence to be addressed by the project team. Using more technical skylights may allow only the sunlight to enter a building rather than additional heat.

The orientation of the building, configuration, the glass types for the windows, the layout of the rooms, the types of furnishings, and the colors of surfaces should all be considered during the design to maximize the use of daylighting without creating any discomfort. Spaces that are designed to efficiently use daylighting as their primary lighting source are called **daylit** spaces.

Following are some benefits of daylighting:

- Positive mental and physical effects on the building occupants
- Increased occupant productivity and comfort
- Cost savings due to the decreased usage of artificial lighting
- Cost savings due to increased occupant productivity

DAYLIGHTING SYSTEMS AND DESIGN

Daylighting systems can be categorized as either passive or active. **Passive daylighting** is a system that collects the sunlight using static and nonmoving items such as windows, glass doors, some skylights, light tubes, and light shelves. **Active daylighting** is a system that tracks and collects the sunlight using mechanical devices; however, these equipments may not function well on cloudy days.

Light tubes, also called solar tubes or sun pipes, are structures that are used to transport sunlight inside a building. To illustrate, imagine a big pipe with a reflective inside surface that goes through the roof to the ceiling of a room. It captures the sunlight on the roof and transports it inside the pipe all the way down to the room.

Light shelves are horizontal, light-reflecting overhangs that are positioned to reflect the daylight into the desired area of the building.

Use of clerestory windows and appropriately sizing and locating all of the windows are important for increasing the amount of daylight introduced into a space. For office buildings, locating private offices toward the building's core and locating cubicles around the perimeter of the building can be effective strategies to bring daylight into a large area. Using **low cubicle partitions** will allow daylight into the core spaces and provide views to the outdoors.

To control glare and avoid direct sunlight reaching a building occupant, adjustable window shades can give occupants control over excessive brightness and glare. The project team should also evaluate the ground surfaces, adjacent buildings, pavement, and objects, as they all can cause glare.

Installing daylight controls is essential in reducing lighting costs. Lighting controls can dim or turn off the electrical lights completely when daylighting is sufficient. Zoning these controls is necessary since the spaces near the windows will have lots of natural light; thus, they can have dimmed artificial lighting. And the spaces away from the perimeter with less daylight can have higher levels of artificial light.

VIEWS

Having quality views in a building is another factor that will increase an occupant's comfort and productivity. Even if the building is in a great location that can provide wonderful views, the design of the building should maximize the use of those views to the building occupants.

Following are some design strategies to maximize the views:

- Using **vision glazing** (a term used for windows that provide exterior views)
- Using glazing without frits, patterns, colors, fibers, or tints
- Using low cubicle partitions or no partitions in open offices
- Adjusting building orientation, site design, facade, and interior layout to maximize views
- For office buildings, locating the private offices closer to the core of the building
- Removal of objects that obstruct views

It is also important to note that design strategies that provide quality views will also assist in increasing the daylighting inside a building; as some strategies are used for both.

ACOUSTICS

Building acoustics are an important part of green building design that will affect an occupant's health, comfort, communication, and productivity. Students in schools can easily be disturbed by outside noise, as well as people working in open office environments. Hospital patients can have even more problems and privacy issues related to poor acoustics.

For that reason, projects should aim for high acoustic quality by controlling key factors such as sound isolation between spaces and HVAC background noise.

LIGHTING CONTROLS

Providing individual lighting controls with adjustable lighting levels will increase occupant comfort and reduce lighting costs. If a big office space does not contain any individual lighting controls, all the lights should be "on," even if there are few people working in the office. Additionally, installing **occupancy sensors** will shut off light when an occupant is away from his or her desk and also reduce lighting usage.

A major aspect of lighting that will affect occupant comfort is lighting quality. The layout and intensity of lighting fixtures should be thoroughly evaluated during the design. Excess brightness and glare are factors that will cause discomfort to building users.

Installing **task lighting** is another important tool that will allow a user to illuminate an area for a specific task.

TEMPERATURE CONTROLS

Every individual has different temperature preferences. A certain temperature in an office building can make some people happy while other people can try to warm themselves up by putting on their sweaters or jackets.

Providing thermal comfort controls for individual spaces can allow building users to set their desired space temperature. Even though this strategy may not work in open offices, installing radiating heat panels under the desks can solve the problem.

ERGONOMICS

Office layouts, individual workstation design, furnishings, and equipment all contribute to the ergonomics of a building. Since every individual has different needs, the project team should consult and analyze occupant needs. Worker education and training should also be provided upon installation of furniture and equipment. Key ergonomic principles to be implemented should include flexibility, versatility, fit, and postural change.

STRATEGIES TO ADDRESS INDOOR ENVIRONMENTAL QUALITY

- Develop and follow a construction indoor air quality management plan during the construction phase to protect the health of both construction workers and future building occupants.
- Use non-emitting or low-emitting construction materials.
- Protect stored materials, building elements, HVAC equipment, and ductwork from exposure to dust and moisture during the construction phase.
- After the completion of construction, conduct an air quality test or a flush-out to ensure sufficient indoor air quality before occupancy.
- During occupancy, set minimum air quality standards, monitor indoor and outdoor air quality, and adjust ventilation rates to provide the right amount of fresh air for health and comfort without overventilating and wasting energy.
- Protect the quality of air entering the building through filtration and proper intake locations.
- Periodically maintain and replace air filters.
- Prohibit smoking inside the building and within 25 feet of building entrances.
- Install entryway systems such as grates, grilles, or mats to prevent contaminants from entering the building.
- Develop and implement a green cleaning policy.
- Maximize daylighting and occupant views.
- Install operable windows where feasible.
- Provide temperature and ventilation controls to occupants.
- Provide lighting controls to occupants.
- Provide ergonomic features.
- Address acoustics to minimize noise disturbances and support occupant comfort.
- Conduct occupant comfort surveys.

INDOOR ENVIRONMENTAL QUALITY PREREQUISITES AND CREDITS

CONSTRUCTION MANAGEMENT—PREREQUISITE

PREREQUISITE SUMMARY

This prerequisite requires projects to implement specific construction management practices during the construction and pre-occupancy phases to maintain strong indoor air quality—for both construction workers and the occupants who will move in after the work is complete.

During construction, dust and pollutants can accumulate indoors, and that contamination can linger long after the job is finished. The goal is simple: reduce exposure to construction dust, toxic substances, and other airborne contaminants. To achieve this goal, the prerequisite adapts practices from the **SMACNA IAQ Guidelines for Occupied Buildings under Construction, 2nd edition, 2007, ANSI/SMACNA 008–2008, Chapter 3**.

Think about HVAC ducts stored in the construction area without dust protection; after installation, they are supposed to provide fresh air to the building occupants. Nevertheless, it is clear that the dusty air they supply will not be healthy. For that reason, the prerequisite emphasizes keeping contaminants out of HVAC systems, avoiding the use of permanently installed equipment when possible, and replacing HVAC filtration media before occupancy. It also includes strict smoking controls such as prohibiting smoking inside the building and within 25 feet (7.5 meters) of entries, openings, and air intakes during construction.

In addition, proper material storage and source control are essential. Many building materials can be soiled or degraded when exposed to jobsite conditions before installation, especially absorptive materials like carpets, ceiling panels, insulation, and similar finishes. Storing these materials in designated areas protected from moisture, dirt, and debris helps prevent contamination and reduces the risk of long-term indoor air quality problems after occupancy.

Construction activities themselves can also generate significant indoor pollutants. Cutting, sawing, sanding, and painting release airborne particles, fumes, and odors that can migrate into adjacent areas and create unintended exposure. To manage these risks, projects must identify hazards early and implement appropriate controls—such as using personal protective equipment and installing temporary barriers to isolate high-emission work areas. Typical isolation strategies include sealing doors and windows, tenting active work zones, and using dust guards or collectors on power tools to capture particles at the source.

In addition, due to global warming, extreme heat has become a major issue, creating much higher health and safety risks for construction workers. For that reason, this prerequisite also includes measures to protect workers from extreme heat. Preventive measures include providing cool, shaded, or air-conditioned areas for rest, implementing required rest breaks, and scheduling labor-intensive activities during cooler parts of the day. The extreme heat requirements are adapted from **OSHA prevention guidance for preventing heat-related illness.**

PREREQUISITE REQUIREMENTS

Develop and implement construction management practices during the construction and pre-occupancy phases to protect indoor air quality and worker health. At a minimum, the plan must address all of the following:

- **No smoking**: Prevent exposure to secondhand smoke by prohibiting smoking on-site during construction, except in clearly designated areas at least **25 feet** (**7.5 meters**) from the building, with visible no-smoking signage in place.
- **Extreme heat protection**: Reduce heat-related illness risk by putting extreme-heat safeguards in place for construction workers (e.g., cooling/rest areas, break protocols, and work planning during peak heat).
- **HVAC protection**: Keep dust and pollutants out of HVAC systems by avoiding use of permanently installed equipment when possible; if use is unavoidable, maintain effective filtration, then replace all filtration media before occupancy and complete testing and balancing with clean filters installed.
- **Source control**: Minimize indoor contamination by keeping pollutant sources out of the building and protecting vulnerable materials. Store absorptive materials (such as carpet, acoustic ceiling panels, fabric wallcoverings, insulation, upholstery, furnishings, and similar finishes) in clean, dry, designated areas protected from moisture and jobsite debris.
- **Pathway interruption**: Prevent contaminants from migrating to other spaces by isolating high-dust or high-emission activities (cutting, sanding, drywall work, VOC-emitting installations, etc.). Use temporary barriers, seal openings, tent work zones as needed, use walk-off mats at entry points, and capture dust at the source with guards and collectors on tools.
- **Housekeeping**: Maintain a clean jobsite through routine cleaning using high-efficiency particulate filtration vacuums and dust-control methods (wetting agents or sweeping compounds) instead of dry sweeping that re-suspends particles.
- **Scheduling**: Sequence construction activities to reduce indoor air quality impacts, and for renovations, coordinate work to minimize disruption and protect occupied areas.

KEY THINGS TO REMEMBER

- To maintain strong indoor air quality for both construction workers and occupants, LEED requires a no-smoking policy, HVAC protection, source control measures, pathway interruption, effective housekeeping, and scheduling construction activities accordingly.
- LEED refers to the SMACNA IAQ Guidelines for Occupied Buildings Under Construction, 2nd edition (2007), ANSI/SMACNA 008–2008, Chapter 3, to reduce indoor environmental quality risks associated with construction and renovation, and to OSHA prevention guidance for preventing heat-related illness to protect construction workers.

- LEED addresses secondhand smoke by prohibiting smoking on-site during construction, except in clearly designated areas at least 25 feet (7.5 meters) from the building, with visible no-smoking signage in place.
- During construction, project teams should first avoid the use of permanently installed equipment when possible. Only if this is not possible should project teams maintain effective filtration, then replace all filtration media before occupancy and complete testing and balancing with clean filters installed.

FUNDAMENTAL AIR QUALITY—PREREQUISITE

PREREQUISITE SUMMARY

Most of us spend most of our time inside buildings, breathing air delivered by HVAC systems rather than fresh outdoor air. Ventilation rates and contamination inside ductwork are important factors to consider because they can cause serious health effects. Properly ventilated spaces always contain fewer contaminants than poorly ventilated areas.

This prerequisite establishes strict indoor air quality standards for ventilation and airflow monitoring to help maintain adequate fresh air. It does so by first investigating regional and local air quality, as indoor air quality will be significantly influenced by outdoor air quality. Then, the prerequisite requires meeting the ventilation rates specified by ASHRAE Standard 62.1-2022, using enhanced filtration media, monitoring outdoor airflow, and installing entryway systems to reduce contaminants tracked into the building. It requires specific MERV ratings for air filters, even though higher-MERV filters can increase energy use. As mentioned, the minimum efficiency reporting value (MERV) rates air filters based on how well they remove particles from the air—the higher the MERV, the more particles are removed.

For residential spaces and healthcare projects, this prerequisite has additional requirements.

PREREQUISITE REQUIREMENTS

1) Investigate Regional and Local Air Quality

Review the project's outdoor air quality in accordance with **ASHRAE Standard 62.1-2022**, Sections 4.1–4.3, to understand regional and local air quality factors that could affect ventilation and filtration design.

AND

2) Ventilation and Filtration Design

Project teams must meet the requirements of **ASHRAE Standard 62.1-2022**, Sections 5 and 6. Project teams can use the ventilation rate procedure, the IAQ procedure, the natural ventilation procedure, or a combination of these methods. These procedures are different ways to show the building gets enough fresh air—either by meeting minimum outdoor-air flow rates, by controlling indoor pollutants to acceptable levels, by using operable openings like windows for natural airflow, or by combining mechanical and natural ventilation.

Where the ventilation rate procedure or IAQ procedure is used, project teams must comply with the following additional provisions:

a) Filtration requirements: Each central HVAC system serving regularly occupied areas must meet one of the following.

- Use HVAC filters rated MERV 13 in accordance with **ASHRAE 52.2-2017**, or
- Use filters rated ISO 16890 ePM1 50% (or equivalent) which is another internationally used way to specify strong particle filtration, or
- Use in-room air-cleaning units, or
- Use air-cleaning systems verified for safety and effectiveness under **ASHRAE 241-2023** (Section 7.4 and Appendix A).

b) Outdoor air measurement requirements: Install outdoor airflow measuring devices for any mechanical system bringing in more than 1,000 cfm (472 L/s) of outdoor air.

For residential spaces, the prerequisite also requires each dwelling unit to conform to the following requirements:

- Design and install a dwelling-unit mechanical ventilation system complying with **ASHRAE 62.2-2022.**
- Kitchens and bathrooms (including half baths) must have local mechanical exhaust per **ASHRAE 62.2-2022** (Sections 5 and 7), exhausted to the outdoors (not to attics or interstitial spaces). If the exhaust hood system is capable of exhausting in excess of 400 cfm (188 L/s), provide makeup air at a rate approximately equal to the exhaust air rate.
- Unvented combustion appliances must not be installed, except ovens and ranges.
- Install a carbon monoxide monitor on each floor of each dwelling unit.
- Fireplaces and wood stoves must have solid glass enclosures/doors that seal when closed.
- Space- and water-heating combustion equipment must be closed combustion (sealed supply air and exhaust) or power-vented, or located in a detached utility building/open-air facility.

AND

3) Entryway System Design

Install permanent entryway systems to capture dirt and particulates at primary exterior entrances. Examples include grilles, grates, slotted systems, and rollout mats.

According to LEED, **primary exterior entrances** are the main entry points with the most visibility and accessibility. These are the doors people naturally walk toward—often highlighted with design features such as canopies or biophilic elements that attract entry. **Nonprimary exterior entrances**, on the other hand, are less visible and used less frequently, such as service access points or lift lobbies, side or back doors, garage and parking level entries, emergency exits, connections between concourses, and atrium entries.

Note that although LEED does not enforce a minimum length requirement for entryway systems, LEED advises project teams to use at least **10 feet (3 meters)** in the primary direction of travel.

KEY THINGS TO REMEMBER

- LEED requires all projects to maintain adequate fresh air by investigating regional and local air quality, increasing ventilation rates, requiring enhanced filtration media, monitoring outdoor airflow, and requiring entryway systems to reduce contaminants tracked into the building.
- LEED refers to the ASHRAE 62.1-2022 and ASHRAE 62.2-2022 to achieve acceptable indoor air quality.
- LEED refers to the ASHRAE 52.2-2017 for air filter standards and requires HVAC filters rated MERV 13.
- LEED refers to the ASHRAE 241-2023 for air-cleaning systems.
- LEED does not enforce a minimum length requirement for entryway systems, but advises project teams to use at least 10 feet (3 meters) in the primary direction of travel.

NO SMOKING OR VEHICLE IDLING—PREREQUISITE

PREREQUISITE SUMMARY

This prerequisite includes strict requirements to prevent smoking both outside and inside the building, thereby improving air quality and protecting nonsmokers from the effects of secondhand smoke.

For schools, the prerequisite restricts smoking on the entire site without any exceptions.

The prerequisite also prohibits vehicle idling, as an idling engine releases pollutants that can worsen asthma, allergies, and heart and lung conditions.

PREREQUISITE REQUIREMENTS

1) Prohibit Smoking

Prohibit smoking inside the building. Also prohibit smoking outside the building except in designated smoking areas located **at least 25 feet (7.5 meters)** from all entries, outdoor air intakes, and operable windows. For schools, no smoking is allowed on the entire site. Project teams should also post no-smoking signage in locations they consider most effective.

For residential projects, these requirements apply to all interior and exterior areas except inside dwelling units and on private balconies. If a residential project allows smoking inside dwelling units, then any unit where smoking is permitted must meet compartmentalization requirements by performing a blower door test in accordance with **ANSI/RESNET/ICC 380** (or equivalent).

AND

2) Prohibit Vehicle Idling

Prohibit vehicle idling on-site and communicate the no-idling policy to occupants. A vehicle idling policy requires occupants not to leave engines running when parked or waiting, unless an exception is truly necessary.

Project teams should also post no-idling signage in locations they consider most effective.

Core and shell projects need to communicate the building's indoor and outdoor smoking and vehicle idling policies in tenant guidelines.

KEY THINGS TO REMEMBER

- LEED requires smoking to be prohibited outside the building except in designated smoking areas located at least 25 feet (7.5 meters) from all entries, outdoor air intakes, and operable windows.
- For schools, LEED restricts smoking on the entire site with no exceptions.
- In residential projects, each residential unit must be compartmentalized to prevent excessive air leakage between units and must meet compartmentalization requirements by performing a blower door test in accordance with ANSI/RESNET/ICC 380 (or equivalent).

ENHANCED AIR QUALITY—CREDIT

- 1 Point (New Construction)

CREDIT SUMMARY

This is the credit of the Fundamental Air Quality prerequisite. It goes beyond the prerequisite by increasing ventilation rates even further to enhance indoor air quality. Providing additional outdoor airflow above the rates defined in the prerequisite will further reduce indoor air pollutants and provide many health benefits for building occupants.

The credit also requires more stringent design limits for ozone, formaldehyde, and PM2.5, which are among the most common and harmful air pollutants that can affect occupant health.

CREDIT REQUIREMENTS

Design the building to exceed the requirements of **ASHRAE Standard 62.1-2022**, Section 6.

Option 1: Increased Ventilation (1 Point)

Increase the breathing-zone outdoor air ventilation rates by at least 15% above the minimum rates established under the Fundamental Air Quality prerequisite. This increased outdoor air must be delivered to at least 95% of all regularly occupied spaces.

OR

Option 2: Enhanced Indoor Air Quality Design (1 Point)

Implement more stringent design limits for ozone, formaldehyde, and PM2.5 defined by LEED.

KEY THINGS TO REMEMBER

- LEED requires smoking to be prohibited outside the building except in designated smoking areas located at least 25 feet (7.5 meters) from all entries, outdoor air intakes, and operable windows.

OCCUPANT EXPERIENCE—CREDIT

- 1–7 Points (New Construction)

CREDIT SUMMARY

As its name implies, this credit awards projects that increase occupant experience through connections to nature, improved comfort, and healthier indoor environments. Projects may pursue a range of strategies, such as incorporating biophilic design and quality views; creating adaptable spaces that give occupants choice and flexibility to meet their needs; and improving thermal, acoustic, and lighting conditions. Teams can select the options that best align with their building type and occupant needs.

CREDIT REQUIREMENTS

Option 1: Biophilic Environment (1–4 points)

Under this option, project teams can earn points by integrating biophilic design in Path 1 and/or providing quality views in Path 2.

Biophilic design must demonstrate engagement with nature, support human well-being, build an emotional connection to place, encourage positive relationships between people and nature, and use integrated design solutions.

For views, project teams must provide outdoor natural or urban views for at least 75% of the regularly occupied floor area through glazing.

AND/OR

Option 2: Adaptable Environment (1 point)

Give occupants choice and flexibility to adjust their spaces to meet their needs. Offer options for thermal comfort, sound, and lighting so occupants can change their environment or move between different sensory settings. Include at least one accessible quiet space.

AND/OR

Option 3: Thermal Environment (1 point)

Design occupied spaces to meet **ASHRAE 55-2023, Thermal Environmental Conditions for Human Occupancy** and consider real-life thermal comfort issues, such as seasonal changes, avoiding overcooling, helping occupants transition between thermal conditions, and supporting different activity levels (including people doing high-movement tasks).

AND/OR

Option 4: Sound Environment (1–2 points)

Early in the design process, define the desired sound environment for each indoor and outdoor space based on its use and occupant needs. Establish acoustic criteria and identify design strategies to meet the acoustic expectations for each space.

AND/OR

Option 5: Lighting Environment (1–6 points)

Under this option, project teams earn points by making lighting more comfortable and useful for occupants. This includes:

- Controlling solar glare by providing manual or automatic glare-control devices in all regularly occupied spaces that receive direct or reflected sun penetration
- Providing high-quality electric lighting
- Improving daylight access by placing at least 30% of regularly occupied spaces near qualifying glazing
- Using a daylight simulation to understand and optimize access to daylight and visual comfort

KEY THINGS TO REMEMBER

- For thermal comfort, LEED refers to ASHRAE 55-2023, Thermal Environmental Conditions for Human Occupancy.
- LEED awards projects that enhance occupant experience by incorporating biophilic design and quality views; creating adaptable spaces that give occupants choice and flexibility to meet their needs; and improving thermal, acoustic, and lighting conditions.

ACCESSIBILITY AND INCLUSION—CREDIT

- 1 Point (New Construction)

CREDIT SUMMARY

This credit focuses on designing a building that works for more people, more of the time. By considering the physical, sensory, and cognitive needs of building occupants, the credit encourages project teams to go beyond basic accessibility measures and create environments that support people with different physical abilities, ages, and needs. This is the idea behind **inclusive design** (also called **universal design**), which aims to make buildings accessible to people of all ages, abilities, and demographics.

To earn the credit, project teams must first comply with all locally applicable accessibility codes identified in the Human Impact Assessment prerequisite (or meet a basic minimum set of accessibility measures if no code exists). Project teams must then go beyond code by selecting and implementing at least 10 additional strategies defined by LEED.

CREDIT REQUIREMENTS

All projects must support access for people with physical disabilities by meeting all locally applicable accessibility codes identified in Human Impact Assessment prerequisite. If no local code exists, the project team must provide ramps where elevation changes occur along accessible routes or at regularly used entrances, ensure doors have at least 32 inches (0.86 meters) of clear width, and ensure that reception desks, security counters, and service counters all have a wheelchair-accessible section with a front approach.

In addition to these requirements, project teams must select and implement at least 10 of the following strategies, which are grouped into four areas: accessibility for physical diversity; accessibility for safety and aging; accessibility for social health; and accessibility for navigation.

Accessibility for physical diversity focuses on making everyday movement easier for people who use mobility devices or need simpler access. Examples include installing hands-free door operators at regularly used entrances, providing meeting spaces that can accommodate mobility devices for at least 10% of occupants, and providing an alternate accessible route whenever stairs or other inaccessible routes are present.

Accessibility for safety and aging focuses on reducing slips, trips, confusion, and injuries. Typical measures include providing nonslip flooring, securing rugs and using transition strips, making full-height glazing visible through markings or railings, providing both audible and visual emergency alerts, and using closed stair risers.

Accessibility for social health supports basic dignity and comfort needs so more people can use the building without barriers. Strategies include providing lactation room pods, adding adult changing stations in accessible restrooms, and supporting neurodivergent users through integrated biophilic design.

Accessibility for navigation helps people understand where they are and how to get where they need to go. Approaches include clear wayfinding that identifies major functions and exits, use of diagrams and symbols, braille and visual/auditory cues, continuous linear indicators along paths of travel, color and pattern cues for key access spaces, and tactile maps for orientation.

RESILIENT SPACES—CREDIT

- 1–2 Points (New Construction)

CREDIT SUMMARY

This credit promotes resilient building design strategies that protect occupant health and well-being during climate-related disruptions and other events that can compromise indoor environmental quality. As climate-related hazards and emergency conditions become more frequent, buildings are increasingly expected to maintain safe, functional, and adaptable indoor environments even during atypical events. This is where passive survivability comes in. **Passive survivability** refers to a building's ability to keep occupants safe by maintaining critical life-support conditions even when services such as power, heating fuel, or water are lost.

The credit can be achieved through multiple pathways that address both mechanical and passive resilience measures. These include providing special operating modes for episodic outdoor pollution events and infection risk reduction, demonstrating thermal habitability during power outages in extreme seasonal conditions, and providing operable windows to improve access to outdoor air when mechanical systems are limited.

CREDIT REQUIREMENTS

Option 1: Management Mode for Episodic Outdoor Ambient Conditions (1 point)

For LEED BD+C: New Construction projects only, design building systems so they can switch into an episodic outdoor event management mode, as described in ASHRAE Guideline 44. This mode should address events that could worsen indoor air quality, such as wildland fire smoke. Include this mode in the design and commissioning documents, and confirm during commissioning that it works as intended.

AND/OR

Option 2: Management Mode for Respiratory Diseases (1 point)

Design occupied spaces with the capability to operate in an infection risk management mode that provides the minimum equivalent clean airflow rates required by ASHRAE 241-2023. Document the mode in the design and commissioning documents and verify during commissioning that the mode is properly implemented.

AND/OR

Option 3: Design for Occupant Thermal Safety During Power Outages (1–2 points)

For extreme heat, demonstrate through thermal modeling that the building can maintain thermally habitable conditions during a two-day power outage under peak summertime conditions of a typical meteorological year (TMY). Identify the thermal safety zones in which those conditions will be maintained.

For extreme cold, demonstrate through thermal modeling or Passive House certification that the building can maintain thermally habitable conditions during a two-day power outage under peak wintertime conditions of a TMY. Identify the thermal safety zones in which those conditions will be maintained.

AND/OR

Option 4: Operable Windows (1–2 points)

Design 50% or 75% of regularly occupied spaces to have operable windows that can provide access to outdoor air during heat waves or localized power outages.

AIR QUALITY TESTING AND MONITORING—CREDIT

- 1–2 Points (New Construction)

CREDIT SUMMARY

This credit awards project teams for assessing indoor air quality after construction through preoccupancy air testing and/or continuous indoor air monitoring.

In Option 1, project teams can earn points by testing indoor air quality after construction and before occupancy for common pollutants such as particulate matter and some inorganic gases, and/or by conducting a screening test for total volatile organic compounds (TVOC). Note that VOC refers to an individual volatile organic compound, while **total volatile organic compounds** (**TVOC**) refers to the combined concentration of many VOCs reported as a single value.

In Option 2, on the other hand, project teams can install continuous indoor air monitors that track key parameters such as carbon dioxide (CO_2), particulate matter (PM2.5), total volatile organic compounds, temperature, and relative humidity.

Particulate matter (**PM**) is a mix of tiny solid particles and liquid droplets floating in the air. Even though some PM can be seen with the naked eye, most of it can only be detected using a microscope. PM can carry all kinds of contaminants such as heavy metals, mold spores, dust mites, combustion by-products, pest droppings, chemicals from pesticides, cleaning products, or building materials.

The smallest particles cause the biggest health threat, as they can travel deep into the lungs. PM2.5 refers to particles that are 2.5 micrometers or smaller, which are linked with serious health effects including heart problems, reduced lung function, and even premature death.

CREDIT REQUIREMENTS

Option 1: Preoccupancy Air Testing (1–2 points)

With the completion of construction and before occupancy, conduct baseline indoor air quality (IAQ) testing under ventilation conditions typical for occupancy. The number of measurements required will vary based on the total occupied floor area.

Project teams can earn points through one or both of the following paths:

> Path 1: Particulate Matter and Inorganic Gases (1 point)
> Test for particulate matter (PM) and inorganic gases specified by LEED (such as carbon monoxide and ozone) and show that concentrations do not exceed the defined limits.

AND/OR

Path 2: Volatile Organic Compounds (1 point)
Perform a screening test for total volatile organic compounds (TVOC) using ISO/EPA methods, then calculate TVOC. If TVOC exceeds the specified limit, investigate potential sources by comparing individual VOC results to health-based limits, correct issues, and retest if needed. Also test the individual VOCs specified by LEED and confirm they do not exceed the defined concentration limits.

AND/OR

Option 2: Continuous Indoor Air Monitoring (1 point)

Install indoor air monitors that measure carbon dioxide (CO_2), particulate matter (PM2.5), total volatile organic compounds, temperature, and relative humidity. Monitors must be located 3–6 feet (1–2 meters) above the floor.

KEY THINGS TO REMEMBER

- To ensure good indoor air quality after the construction period, LEED awards projects that conduct baseline indoor air quality testing and/or install indoor air monitors that can continuously monitor indoor air quality.
- Particulate matter (PM) can carry all kinds of contaminants such as heavy metals, mold spores, dust mites, combustion by-products, pest droppings, and chemicals from pesticides, cleaning products, or building materials, and it can even form new harmful compounds through chemical reactions. The smallest particles, such as PM2.5, pose the biggest health threat, as they can travel deep into the lungs.

END OF CHAPTER EXERCISES

EXERCISE: MATCHING GAME

Match the following terms to their definition.

Terms		
Source Control	Demand-Controlled Ventilation	Active Daylighting
Inherently Nonemitting Materials	Minimum Efficiency Reporting Value (MERV)	
Flush-Out	Light Shelves	

1) ____________________________: A ventilation strategy that automatically adjusts outdoor airflow in real time using sensor signals.

2) ____________________________: A rating system that measures how effectively an air filter removes particles from the air.

3) ____________________________: A system that tracks and collects the sunlight using mechanical devices; however, these equipments may not function well on cloudy days.

4) ____________________________: The practice of identifying sources of indoor contaminants and taking measures to prevent them from reaching the indoor environment.

5) ____________________________: Materials with very low or no VOC content.

6) ____________________________: The process of supplying large volumes of fresh air to the building before or during occupancy to remove contaminated air and establish the desired level of indoor air quality.

7) ____________________________: Horizontal, light-reflecting overhangs that are positioned to reflect the daylight into the desired area of the building.

ANSWER: MATCHING GAME

1) Demand-Controlled Ventilation	4) Source Control	7) Light Shelves
2) Minimum Efficiency Reporting Value (MERV)	5) Inherently Nonemitting Materials	
3) Active Daylighting	6) Flush-Out	

PRACTICE EXAM

To access the end-of-chapter practice exam online in our exam simulator, scan the QR code below with your laptop, phone, tablet, or another compatible device, then log in using your email address and password. If you have trouble scanning the QR code, visit www.projectific.com/account and log in using your email address and password. If you have not yet created a password, please refer to the *How to Access the Exam Simulator* section at the back of the book for instructions.

CHAPTER 11: PROJECT PRIORITIES

LEED projects span the entire globe. They exist in different climate zones, typologies, and population densities, and they face varying levels of water and energy availability—as well as different local challenges. On top of that, every project is unique: it serves different users, responds to different goals, and operates under different constraints. Therefore, not every project should chase the same sustainability checklist. A strategy that is critical in a water-stressed region may be far less meaningful where water is abundant. Similarly, what matters most for a hospital will not be the same as what matters most for a warehouse or a residential tower. This credit category was created to give projects more flexibility to pursue what matters most for their specific context and priorities, and to encourage teams to innovate and accelerate new sustainable building strategies.

Moreover, industry knowledge and technologies evolve quickly. In previous LEED versions, project teams had to wait for the next major rating system release before new ideas, metrics, and solutions could be formally recognized. With LEED v5, USGBC created the **Project Priorities Library**, where new credits can be added as they are developed. In other words, LEED no longer has to wait for the next major version release to incorporate emerging best practices. If the industry learns something important, or if a better strategy becomes available, the system can respond between major rating system updates.

Under this credit category, project teams can earn up to 9 points by selecting credits from the USGBC's Project Priorities Library. These credits are designed to recognize efforts that are especially meaningful based on location, project type, performance ambition, or innovation. There are five credit types under the USGBC's Project Priorities Library as follows:

- Regional priority
- Project-type credits
- Exemplary performance
- Pilot credits
- Innovation strategies

Let's take a closer look at each type of credit.

1) Regional priority:

The **Regional Priority credit** is designed to reward projects for focusing on what matters most in their specific location. To earn points under this credit, the project team should identify a geographically specific environmental, social equity, or public health priority for the project site and then demonstrate achievement of a credit from the main LEED credit categories that addresses that priority in a significant, measurable way. For example, if the project is in a water-critical location, the project team may target the Enhanced Water Efficiency credit and make substantial, measurable reductions in potable water use to directly address local water scarcity. In this case, in addition to receiving points from the credit, the project may earn an extra point by submitting it as a Regional Priority credit.

2) Project-type credits:

Project-type credits are designated by USGBC to address the specific, unique needs of a particular adaptation or building application.

For example, the Places of Respite credit is a project-type credit available only to healthcare projects. It aims to provide patients, staff, and visitors with the health benefits of the natural environment by creating outdoor places of respite on the healthcare campus. To meet the requirements, projects must provide respite areas accessible to patients and visitors equal to 5% of the building's net usable program area, plus additional dedicated respite areas for staff equal to 2%.

3) Exemplary performance:

For eligible LEED credits, performance beyond the LEED credit requirements can earn projects **exemplary performance points**, which are generally achieved by doubling the credit requirements or reaching the next incremental percentage threshold.

For example, in the Electric Vehicles credit, projects can earn up to 2 points by installing electric vehicle supply equipment that meets the credit's criteria. If the project team doubles the highest electric vehicle supply equipment parking-point threshold, they will receive 1 additional point for achieving a level of exemplary performance.

Another example is the Enhanced Water Efficiency credit. In Option 1 of this credit, project teams can earn 8 points if they reduce the project's overall potable water use by 60% through conservation and the use of alternative water sources. If they go beyond that and achieve a 65% reduction—which is the exemplary performance threshold for that credit—an extra point will be awarded.

Don't forget that not all LEED credits are eligible for exemplary performance; moreover, eligible credits have different exemplary performance requirements.

4) Pilot credits:

As discussed in the introductory chapters, pilot credits are tested across all rating system types and credit categories and include credits proposed for the next version of LEED. Project teams can pursue any of these pilot credits under Project Priorities credit and earn points by providing USGBC with feedback on the credit. Based on the feedback USGBC receives from projects that pursue these credits, pilot credits may become official credits in the next LEED rating system.

5) Innovative strategy:

LEED encourages projects to go beyond established green building principles by implementing new and unique strategies that are not currently addressed in the LEED rating systems. If projects develop innovations that are not covered in the rating system and can demonstrate reduced environmental impacts, increased decarbonization, and improved social impacts, they may receive additional points under the Project Priorities credit. Alternatively, they may also choose to achieve innovation credits from the USGBC's Project Priority Library.

PROJECT PRIORITIES CREDITS

PROJECT PRIORITIES—CREDIT

- 1–9 Points (New Construction)

CREDIT SUMMARY

This credit awards up to 9 points for pursuing any combination of the following pathways in the Project Priority Library: Regional Priority, Project-Type Credits, Exemplary Performance, Pilot Credits, and Innovative Strategies.

CREDIT REQUIREMENTS

Project teams can earn a maximum of 9 points by achieving any combination of the following:

1) Regional Priority

Achieve the Regional Priority credit from the Project Priority Library. When identifying the regional priorities for the project site, the team must support the selection with background analysis—such as the Climate and Natural Hazard assessment or Human Impact Assessment completed under the Integrative Process prerequisites, local weather maps, government studies, regional climate action plans, or assessments from organizations like the IPCC or WRI—and explain how the adopted strategy responds to the priority.

AND/OR

2) Project-Type Credits

Achieve a project-type credit from the Project Priority Library.

AND/OR

3) Exemplary Performance

Achieve an exemplary performance credit from the Project Priority Library.

AND/OR

4) Pilot Credits

Achieve a pilot credit from the Project Priority Library.

AND/OR

5) Innovative Strategies

Achieve an innovation credit from the Project Priority Library, or demonstrate significant, measurable environmental performance using a strategy that is not addressed in the LEED v5 rating system. To use this pathway, the project team must identify all of the following:

- The intent of the proposed innovation credit
- Proposed requirements for compliance
- Proposed submittals to demonstrate compliance
- The design approach or strategies used to meet the requirements

KEY THINGS TO REMEMBER

- Know the purpose of the Regional Priority credit, Project-Type credits, Exemplary Performance credits, Pilot credits, and Innovation credits.
- The Project Priorities Library includes Regional Priority, Project-Type Credits, Exemplary Performance, Pilot Credits, and Innovation credits, and awards up to 9 points for pursuing any combination of them.
- When identifying the regional priorities for the project site, the team must support the selection with background analysis—such as the Climate and Natural Hazard assessment or Human Impact Assessment completed under the Integrative Process prerequisites—and explain how the adopted strategy responds to the priority.

LEED AP—CREDIT

- 1 Point (New Construction)

CREDIT SUMMARY

This credit awards 1 point for having a LEED AP with a Building Design and Construction (BD+C) specialty as one of the principal participants on the design team. Legacy LEED APs (LEED APs without a specialty) do not qualify for this credit.

The selected LEED AP with specialty must hold an active credential at the time of the project's certification review.

CREDIT REQUIREMENTS

At least one principal participant on the design team must be a LEED AP with a BD+C specialty. A project cannot earn this credit if it employs a LEED AP O+M as a principal participant for a LEED BD+C project.

KEY THINGS TO REMEMBER

- Legacy LEED APs (LEED APs without a specialty) do not qualify in this credit.
- Projects can receive only 1 point under this credit, even if they employ more than one LEED AP with a BD+C specialty.
- With this credit, the Project Priorities credit category provides a total of 10 points (the Project Priorities credit awards 9 points, and this credit awards 1 point).

PRACTICE EXAM

To access the end-of-chapter practice exam online in our exam simulator, scan the QR code below with your laptop, phone, tablet, or another compatible device, then log in using your email address and password. If you have trouble scanning the QR code, visit www.projectific.com/account and log in using your email address and password. If you have not yet created a password, please refer to the *How to Access the Exam Simulator* section at the back of the book for instructions.

CHAPTER 12:

LEED® GREEN ASSOCIATE™ EXAM

Congratulations on nearing the end of your journey toward earning your certification! In this last chapter, we will review the exam content, present final exam tips, and explain how to determine your readiness level for the exam. We will end this chapter by detailing the exam policies, procedures, and other important information you should know before taking the exam.

EXAM CONTENT

The LEED Green Associate exam contains 100 randomly delivered multiple-choice questions and must be completed in 2 hours. The exam questions reflect Knowledge Domains aligned with the LEED v5 rating system credit categories, which we have covered in depth throughout this book.

Below is the breakdown of the exam content areas and the number of questions per category:

- LEED Process (15 questions): USGBC/GBCI fundamentals, LEED rating system structure, certification benefits, available resources, rating system families, development process, category goals, impact areas, and certification process.
- Integrative Process Planning and Assessments (6 questions): Early assessment, systems thinking, life-cycle cost analysis, and roles of integrative project team members.
- Location and Transportation (9 questions): Site selection, transportation demand management, active and alternative transportation, VMT reduction, and equitable development.
- Sustainable Sites (8 questions): Site assessment, environmental considerations, open space, habitat conservation, heat island reduction, light pollution reduction, and rainwater management.
- Water Efficiency (10 questions): Outdoor and indoor water use reduction, graywater and rainwater use, smart irrigation, efficient fixtures and appliances, leak detection, and water metering.
- Energy and Atmosphere (15 questions): Building loads, energy efficiency, energy modeling, passive strategies, renewable energy, green power, energy monitoring, commissioning, and decarbonization.
- Materials and Resources (11 questions): Life-cycle impacts, embodied carbon, material selection, product declarations, VOCs, reuse, recycling, repurposing, and waste management.
- Indoor Environmental Quality (11 questions): Indoor air quality, low-emitting materials, ventilation, lighting, acoustics, thermal comfort, biophilic design, accessibility, and inclusion.

FINAL EXAM TIPS

In this section, we will present essential test-taking tips and strategies that are vital for exam success.

FINAL EXAM TIPS

- **Read carefully. Make sure you read all the questions and the choices very carefully. If a question seems to have more than one answer, make sure you thoroughly understand the question and pay special attention to the wording. It doesn't matter how well you know the exam content; if you don't read carefully, the exam can easily trick you into selecting the wrong answer.**
- **Passing score.** The LEED Green Associate exam is scored between 125 and 200. A score of **170 or higher is required to pass**. However, the exam contains both scored and unscored questions. Unscored questions are used by GBCI to collect performance data and potentially serve as future exam questions. Test-takers will not know which questions are scored and which are unscored. Because of this, there is no way to know exactly how many correct answers you need in order to pass.

- **Time management.** During the exam, you will have **2 hours** to answer **100 questions**, which means you have about **72 seconds on average per question**. If you are stuck between two very close answers, instead of spending significant extra time trying to decide, mark the question and come back to it later. Often you will find it much easier to answer such questions when you return to them with a fresh perspective.
- **The test itself may help you.** If you forget something and struggle to answer a question, note that questions or answer choices you see later in the exam can help you remember the right answer to a previous question.
- **Answer every single question.** Leaving a question blank will not award any points, whereas making your best guess may result in the correct answer.
- **Test early in the day.** Schedule your exam for the morning and get a good night's sleep the night before. The exam requires strong concentration, and taking it with a fresh mind is the easiest way to boost your score. Don't study on exam day — take the test with a clear and refreshed mind.
- **Be prepared for distractions.** If you are taking your exam at a testing center, be prepared for potential noise distractions. Even though most settings will be quiet, some may not be. These centers offer a variety of exams, so expect to see people entering and exiting the testing area throughout your exam.
- **Manage test anxiety.** If you feel anxiety during the exam, pause for a moment. Take a deep breath, hold it for five seconds, and then slowly release it. Repeat this process a few times if needed. Thinking about a very funny moment in your life also works well to de-stress before moving on. Without managing your anxiety, your mind will struggle to function effectively, making every question seem more difficult than it really is. If a challenging question heightens your anxiety, use this breathing exercise, flag the question for later review, and move on. You will often find that when you return to that question later, it will be much easier to answer.
- **Use all your time.** If you have extra time at the end of the exam, use it to review your answers. Start with the questions you flagged and those you feel most unsure about.

NO NEED TO BUY ADDITIONAL STUDY MATERIALS

By thoroughly studying this book, you can pass the exam with flying colors.

HOW TO KNOW YOU'RE READY

As mentioned at the beginning of our journey, this book offers a streamlined system to minimize the time you spend reviewing the entire content and significantly shorten your overall study time. Upon completing this chapter, the *Key Things to Remember* sections will enable you to review the key points you need to know for the exam in no time. Additionally, all key terms and concepts are written in bold so you can easily find and review them after you finish the book.

Once you complete this chapter, you should review the entire content by returning to the beginning of the book and going through all the *Key Things to Remember* sections and the terms and concepts highlighted in bold. If any exam tip, term, or concept seems unclear, reread the relevant section to reinforce your understanding. Additionally, the green building strategies covered in each credit category are summarized in the *Summary of Strategies* sections in Chapters 5–11. Appendices A and B also summarize the key standards and programs you need to know for the exam.

At this point, you are ready to take the final full-length practice exam in the exam simulator. If you haven't created your account for accessing the exam simulator, you can find instructions on how to access it in the *How to Access the Exam Simulator* section at the back of the book. As you complete the practice exam, note the topics in which you performed poorly and revisit those sections in the book to reinforce your understanding.

To summarize, before the exam, make sure you feel comfortable with:

- Everything mentioned under Key Things to Remember sections at the end of every prerequisite/credit
- The summary of strategies for each credit category in Chapters 5–11
- Appendix A, which summarizes the ASHRAE standards
- Appendix B, which summarizes other important standards and programs

If you have thoroughly studied this book and feel comfortable with the items listed above, you can confidently take the actual exam.

LEED GREEN ASSOCIATE EXAM POLICIES AND PROCEDURES

In this section, we will provide details on the LEED Green Associate exam policies, procedures, and other important aspects you should know before taking your exam. However, as GBCI's policies and procedures are subject to change, it is also recommended that you visit USGBC's website and download the latest LEED Green Associate Candidate Handbook.

HOW TO REGISTER FOR THE EXAM?

To register for the LEED Green Associate exam, you need to complete the following steps:

1) Create a USGBC site user account at www.usgbc.org/credentials, or log in if you already have one. **When filling out the required information, make sure that the name you enter exactly matches your name on the ID you will present to take the exam**. Any discrepancy may disqualify you from taking the exam.

2) Under the *Credentials* section, select the LEED Green Associate credential and follow the instructions to complete the application. Full-time and part-time students can take advantage of discounted rates for the LEED Green Associate exam. When submitting your exam registration, you will be required to enter your school, student ID number, and graduation date to receive the student discount.

3) Once your application is submitted and payment is processed, you will receive an Eligibility ID (EID) and will be redirected to prometric.com/gbci to schedule your exam date and location. Alternatively, you may choose to take the exam online through remote proctoring. Your EID is valid for 12 months from the date of registration.

4) When the exam is scheduled, you will receive a confirmation number onscreen and an email through Prometric. Keep your confirmation number, as you will need it to confirm, cancel, or reschedule your appointment.

5) Once you register and pay for your exam, you have one year to schedule and take your exam session. You may request one six-month extension of this one-year period.

WHERE TO TAKE THE EXAM?

The LEED Green Associate exam can be taken at Prometric testing centers around the world, or you can take the exam online, in the comfort of your home or office, while being monitored by an offsite proctor. In the following sections, we will discuss the policies and procedures for both the testing center and the online proctored test.

TESTING CENTER POLICIES AND PROCEDURES

Test-takers who choose to take their exam at a Prometric testing center should arrive 30 minutes before their scheduled appointment to sign in and present their identification, Eligibility ID (EID), and the confirmation number, which are emailed upon scheduling the appointment. Test-takers who arrive later than their scheduled exam time will lose their seat.

When signing in, test-takers must present a valid, non-expired, government-issued photo ID that also contains a signature. Once more, your name on the identification must exactly match the name on your exam registration. In case of a discrepancy, you may not be permitted to take your exam.

Nothing can be brought to the testing area. Small lockers may be available at the testing center to store your wallet, cell phone, keys, and other small items. However, laptops, briefcases, or large purses should not be taken to the testing center.

Throughout the exam session, you should remain seated. If you experience an issue or need to take a break, raise your hand to notify test center staff. Note that the exam will not pause if you take a break, and you must get the test proctor's permission to leave the room.

Expect to see other individuals entering and exiting the testing area throughout your exam, as Prometric testing centers offer a variety of exams. If the testing center is in a busy area, be prepared for potential noise distractions.

POLICIES AND PROCEDURES FOR ONLINE PROCTORED TEST

To take the LEED Green Associate exam online, you will need a computer equipped with a webcam, a stable internet connection, and a private, quiet space where no one will enter or pass through your testing area.

Before scheduling your exam, be sure to review the online proctoring requirements on the Prometric website at prometric.com/gbci. On that website, you will have the opportunity to run a system test to ensure your computer and internet connection meets all the necessary criteria for a smooth testing experience. Taking this step before scheduling your exam is crucial to avoid any technical issues that could interrupt or invalidate your exam session. If your system cannot pass the test, you may instead choose to take the exam at a testing center.

On exam day, you will need to check in for the online proctored test before your scheduled appointment. During the check-in process, you will need to run a system check and provide photos of yourself, your identification, and your testing space. Your identification must be a valid, non-expired, government-issued photo ID that also contains a signature.

Once you submit your photos, they will be reviewed by the offsite proctor. If there is any issue, the proctor will contact you. To speed up the check-in process, make sure that your desk and the area around you are clear of all personal and preparatory items, including books, notes, and other materials. Otherwise, the proctor may ask you to show all sides of your testing environment via your webcam.

Throughout your online exam session, you will be under surveillance via your webcam and microphone by the offsite proctor. If someone enters your testing space or if there are loud distractions, you risk having your exam terminated. For this reason, choose a testing space free of interruptions and background noise.

Once everything meets the requirements, the proctor will start your exam. You should remain seated in front of your webcam throughout the exam session. If you encounter any technical issue during your exam, the proctor will attempt to resolve it in real time.

SPECIAL TESTING ACCOMODATIONS

Accommodations can be requested if you have a documented disability that would prevent you from taking the exam under normal testing conditions. Prometric complies with the provisions of the Americans with Disabilities Act (ADA). There is no additional charge for special accommodations.

DURING THE EXAM

EXAM TUTORIAL AND EXIT SURVEY

Before the exam, there will be a 10-minute tutorial to demonstrate the exam software. At the end of the exam, there will be a 10-minute optional exit survey. The time allocated for the tutorial and the survey is not included in your 2-hour examination time.

THE EXAM

The LEED Green Associate exam contains 100 randomly delivered multiple choice questions that must be completed in 2 hours.

As the exam starts, you will be presented with one question at a time. For every question, you have the option to answer the question, answer and mark it for later review, or skip it. You have the flexibility to move back and forth among the questions at any point and change your answer to any question you have previously addressed. The exam software will also list all the questions you have either skipped or marked for review so that you can easily revisit them.

Make sure you read all the questions and the choices very carefully. If a question seems to have more than one answer, make sure you thoroughly understand the question and pay special attention to the wording. It doesn't matter how well you know the exam content; if you don't read carefully, the exam can easily trick you into selecting the wrong answer.

Answer every single question since leaving a question blank will not award any points. If you are unsure about the answer to a question and are likely to spend extra time deciding on the correct answer, mark the question and come back to it later. Often, you will find it easier to answer such questions upon returning to them with a fresh perspective. If you forget something and struggle to answer a question, the questions or answer choices you see later in the exam can help you remember the right answer to a previous question.

At the end of the exam, make sure you use any extra time to review your answers, starting with the questions you felt most unsure about.

AFTER THE EXAM

EXAM RESULTS

After the exam, your exam score will be displayed on screen, and a score report will be emailed to you following your exam session. Within 72 hours of your appointment, your exam results will be processed and your Credentials Dashboard will be updated.

Note that candidates who take the LEED v5 beta exam (April–August 2026) will not receive immediate results; beta exam scores are expected to be released in October 2026.

PASSING THE EXAM

As soon as you have passed the exam, you can use the title "LEED Green Associate" and/or the logo. Please remember that the abbreviation "LEED GA" is not approved by USGBC and should not be used.

If a LEED AP with specialty credential is earned after becoming a LEED Green Associate, your LEED Green Associate credential will be replaced by the LEED AP with specialty credential.

FAILING THE EXAM

The full exam fee will be charged for each scheduled exam session after failing an exam. After three unsuccessful attempts, you must wait 90 days before submitting a new registration to GBCI. With thorough preparation, however, failing is not an option and you can confidently pass the exam with flying colors on your first-try.

HOW TO MAINTAIN YOUR CERTIFICATION

All LEED Green Associates must earn 15 continuing education (CE) hours within 2 years of earning their credential, while all LEED APs with specialty must earn 30 continuing education (CE) hours within 2 years of earning their credential.

CONCLUSION

First of all, I would like to congratulate you on finishing this book and personally thank you for choosing this guide for your LEED Green Associate exam preparation. My primary goal in writing this book was to create a single source that covers the complete exam scope and simplify green building concepts to equip readers to successfully pass the exam and effectively apply this knowledge in real life. I truly hope this book has exceeded your expectations.

To truly evaluate and sustain our performance, we all need feedback. With that in mind, please share any comments about this book, your exam score, and/or any other opinions you would like to share with me. If you found this book helpful, spreading the word by leaving a review on your favorite book review site, such as Amazon, Goodreads, or others, would mean the world to me. Your review supports my work and helps guide other readers.

If you'd like to receive my latest articles, books, and courses, be sure to join my mailing list.

Thank you for being part of this journey. I wish you all the best in your exam and your career!

Scan the following QR codes to:	
Send your feedback to the author and/or subscribe to the author's mailing list	**Follow the author on LinkedIn**

APPENDIX A – SUMMARY OF ASHRAE STANDARDS

Name of the Standard / Program	Keywords	Related Prerequisites / Credits
ASHRAE 52.2-2017	Filtration requirements	Fundamental Air Quality Prerequisite
ASHRAE 55-2023	Thermal comfort requirements	Occupant Experience Credit
ASHRAE Standard 62.1-2022	Ventilation and outdoor air requirements	Fundamental Air Quality Prerequisite; Enhanced Air Quality Credit
ASHRAE 62.2-2022	Residential ventilation and exhaust requirements	Fundamental Air Quality Prerequisite
ANSI/ASHRAE/IES Standard 90.1 / ASHRAE 90.1-2022	Energy efficiency, commissioning, energy metering, thermal bridging, peak load reduction, energy simulation, and energy performance improvement	Minimum Energy Efficiency Prerequisite; Fundamental Commissioning Prerequisite; Energy Metering and Reporting Prerequisite; Reduce Peak Thermal Loads Credit; Enhanced Energy Efficiency Credit
ASHRAE Standard 202-2024	Enhanced commissioning framework	Enhanced Commissioning Credit
ASHRAE 241-2023	Air cleaning and infection risk management	Fundamental Air Quality Prerequisite; Resilient Spaces Credit

APPENDIX B – IMPORTANT STANDARDS AND PROGRAMS

Name of the Standard / Program	Keywords	Related Prerequisites / Credits
Intergovernmental Panel on Climate Change (IPCC)	Referenced for climate resilience assessment	Climate Resilience Assessment Prerequisite
ANSI Consensus National Standard Guide 2.0	Referenced for integrative process guidance	Integrative Design Process Credit
U.S. Department of Agriculture	Defines prime farmlands	Sensitive Land Protection Credit
Federal Emergency Management Agency (FEMA)	Designates flood hazard areas	Sensitive Land Protection Credit
U.S. Endangered Species Act	Defines endangered and threatened species	Sensitive Land Protection Credit
NatureServe	Defines and assigns conservation status ranks for species and ecosystems	Sensitive Land Protection Credit
International Union for Conservation of Nature (IUCN) Red List	Identifies endangered and threatened species	Sensitive Land Protection Credit
Institute of Transportation Engineers Parking Generation Manual	Provides base parking capacity ratios	Transportation Demand Management Credit
ENERGY STAR	Sets EVSE and energy-efficient equipment requirements	Electric Vehicles Credit, Minimum Water Efficiency Prerequisite, Enhanced Energy Efficiency
U.S. EPA 2022 Construction General Permit (CGP)	Sets requirements for the erosion and sedimentation control (ESC) plan	Minimized Site Disturbance Prerequisite

Name of the Standard / Program	Keywords	Related Prerequisites / Credits
American Bird Conservancy Threat Factor Database	Defines bird-friendly glass requirements	Biodiverse Habitat Credit
FORTIFIED Commercial High Wind and Hail Specific Design Requirements	Sets requirements for hail and wind resilient design	Enhanced Resilient Site Design Credit
U.S. National Tsunami Hazard Mitigation Program	Sets requirements for tsunami-resilient design	Enhanced Resilient Site Design Credit
National Wildfire Coordinating Group Standards for Mitigation	Guides wildfire protection measures	Enhanced Resilient Site Design Credit
American Forests	Created the Tree Equity Score	Heat Island Reduction Credit
Illuminating Engineering Society of North America (IESNA)	Develops lighting specifications	Light Pollution Reduction Credit
WaterSense	A program developed by EPA to identify high-performance, water-efficient fixtures and fittings	Minimum Water Efficiency Prerequisite
Green-e	The leading certification program for the green power generation in the United States	Renewable Energy Credit
California Department of Public Health (CDPH) Standard	Referenced for VOC emissions evaluation criteria	Low-Emitting Materials Credit
Sheet Metal and Air Conditioning National Contractors Association (SMACNA) IAQ Guidelines	Describes the necessary control measures to be taken during construction to protect indoor air quality	Construction Management Prerequisite

FINAL PRACTICE EXAM (100 QUESTIONS)

To help you become familiar with the format of the actual LEED Green Associate exam, the full-length practice exam is available through our online exam simulator. To access the full-length practice exam, please refer to the *How to Access the Exam Simulator* section at the back of the book for instructions on creating your password. Once your password has been created, you can log in to the e-learning platform and take the full-length practice exam at any time by scanning the QR code below with your laptop, phone, tablet, or another compatible device, then entering your email address and password.

If you have trouble scanning the QR code, visit www.projectific.com/account and log in using your email address and password. If you need any assistance, please email us at customercare@projectific.com.

This final exam, consisting of 100 questions, is designed to provide you with a real exam simulation to test and reinforce your knowledge, while also identifying weak points that need improvement.

Before starting your exam, our exam simulator will ask you to choose from three different test-taking methods. These methods determine when the exam simulator will score your answers and display the answer explanations. In one option, the exam simulator will display the correct answer and its explanation only after each incorrect response. For each correct response, it will directly proceed to the next question. In another option, the exam simulator will score your answers and display all answer explanations upon completion of the exam. In the third option, the simulator will score your answer and display the answer explanation immediately after you answer each question, regardless of whether your answer is correct or not.

In addition to having a thorough understanding of all exam topics, time management is also important for exam success. Therefore, it is strongly recommended that you take this practice exam in one sitting, within the 120-minute limit, to become accustomed to the timing.

On average, you have 72 seconds to answer each question. When taking this practice exam, and the actual exam, it is important to watch your time to ensure you don't spend too much time on any question. If you are unsure about the answer to a question and are likely to spend a considerable amount of extra time deciding on the correct answer, skip the question and come back to it at the end of the exam. Both our exam simulator and the actual LEED exam simulator will let you skip questions and come back to review them later. You can also change the answer to any test question before the completion of the exam.

GLOSSARY

40/60 Rule: A method used to choose the appropriate rating system for the project if the project seems to fit under multiple rating systems.

Active Daylighting: A system that tracks and collects sunlight using mechanical devices.

Active Transportation: Human-powered mobility such as biking, walking, or wheeling.

Adapted Plants: Types of plants that do not occur naturally in a specific location; however, they can nonetheless adapt easily to the climate of the region.

Adaptive Reuse (**Designing for Flexibility**): A design strategy that allows buildings to accommodate future changes in layouts or floor plans without major reconstruction, particularly useful in buildings that may need frequent modifications.

Addenda: The overarching term that encompasses all changes, improvements, and modifications made to a LEED rating system. Addenda includes four types of changes: corrections, clarifications, LEED interpretations, and alternative compliance paths (ACPs).

Advisory Task Forces: Short-duration groups designated by USGBC staff to informally advise on issues limited in scope during the LEED development process.

Albedo: A measure of surface reflectivity on a scale from 0 to 1, where 0 represents a black surface that absorbs all solar radiation and 1 represents a white surface that reflects all solar radiation.

Alternative Compliance Paths (ACPs): Provide additional options for meeting credit requirements for international projects that need to reference local or regional standards instead of United States–based standards. There are two types of ACPs: Regional ACPs, which are credit requirements that may optionally be applied in specific regions of the world, and pilot ACPs, which are paths USGBC is considering for inclusion in future versions of LEED and can optionally be used in place of existing LEED credits.

Alternative Fuel Vehicles: Vehicles that consume nongasoline, low-polluting fuels like hydrogen, electricity, propane, compressed natural gas, liquid natural gas, methanol, or ethanol.

Alternative Fuel: Low-polluting fuels like hydrogen, electricity, propane, compressed natural gas, liquid natural gas, methanol, or ethanol.

Alternative Water Source: Nonpotable water from on-site surfaces, or freshwater sources, such as graywater, on-site reclaimed water, collected rainwater, captured condensate, and rejected water from reverse osmosis systems. Water from public utilities is excluded.

Arc: The online platform where project teams register their projects for LEED certification, complete prerequisite and credit documentation, and submit applications to GBCI for certification review. It serves as the primary tool for managing the entire LEED certification process from start to finish.

Basis of Design (BOD): Describes the information necessary to accomplish the owner's project requirements, including system requirements, design criteria, standards, guidelines, etc., developed by the architect/engineer.

Bio-Based Materials: Products other than food that are biological products, renewable agricultural materials, or forestry materials derived from biomass. Plants and animals can be examples of bio-based materials; however, hide products such as leather and other animal skin materials are excluded in LEED calculations.

Biofuel: Fuels produced from organic material. Biofuel includes untreated wood waste, landfill gas, agricultural crops or waste, animal waste, and other types of organic waste.

Bioswale: A stormwater control feature that uses a combination of an engineered basin, soils, and vegetation.

Blackwater: Used water that has come into contact with waste. Thus, the water collected from the urinals and toilets can be classified as blackwater.

Blowdown: Removal of the cooling tower's water in order to minimize deposit of scales.

Brownfield Site: Previously developed sites that were contaminated with waste or pollution. A site that has an abandoned building with unknown contamination can also be classified as a brownfield site.

BUG Rating Method: A luminaire classification system that classifies a luminaire according to backlight, uplight, and glare.

Building Automation System (BAS): A computer-based monitoring system which can monitor, coordinate and control every individual building system.

Building Exterior: Defined as everything from the waterproofing membrane, inclusive of the waterproofing membrane.

Building Footprint: Describes the area that the building sits on.

Building Interior: Defined as everything inside the waterproofing membrane.

Built Environment: All the man-made surroundings that are needed for human activity, from roads, to buildings, to neighborhoods.

Charrettes: Intense workshops that are generally held at the beginning of the project and during project milestones. These workshops bring together the entire project team, facility manager, and all the stakeholders and outside experts/consultants. Charrettes allow brainstorming and collaboration among different disciplines and experiences.

Clean Waste: Materials that are left over from construction and demolition that are nonhazardous.

Climate Risk: The likelihood and potential severity of harm resulting from a climate-related hazard, taking into account the building's vulnerability, its occupants, and the surrounding context.

Climate Risk Assessment: A systematic evaluation of how identified climate hazards could affect a building's performance, safety, occupants, and the owner's investment over the project's expected service life.

Closed System: A system that does not produce any waste product by circulating the same medium continuously.

Commingled Waste: Single-streamed waste for recycling.

Commissioning (Cx): A systematic process carried out by skilled personnel to verify that a building's systems are designed, installed, and operating in accordance with the owner's project requirements (OPR), project goals, and design specifications. Building commissioning ensures that the project is designed and built as truly intended.

Compact Development: Development strategies that promote location-efficient neighborhoods with high development density, reducing greenhouse gas emissions and urban sprawl.

Compressed Workweek: A work arrangement in which employees work longer hours on fewer days in exchange for an additional day off during the workweek, reducing commuting frequency and associated fossil fuel usage.

Consensus Committees: Established by the USGBC Board of Directors to support the creation and updating of specific LEED rating systems. Each consensus committee is structured to represent different technical and market interests, and no single interest group may make up more than 25% of the voting membership.

Conventional Irrigation: Common system used for irrigation, such as irrigation through sprinkler heads above the ground.

Cradle-to-Cradle: The practice of designing materials to have "infinite" life cycles by recovering and recycling them into new products, creating a closed-loop system instead of waste.

Cradle-to-Grave: A life-cycle approach that evaluates a product from raw material extraction through manufacturing and use, all the way to final disposal.

Cradle-to-Gate Assessment: A partial life-cycle assessment that evaluates impacts from raw material extraction/harvesting through manufacturing, ending when the product leaves the factory gate (before transportation to the site, use, and end-of-life).

Decarbonization: The process of reducing or eliminating carbon emissions from building operations, embodied materials, refrigerants, and transportation. Decarbonization is the largest impact area in LEED v5, accounting for approximately 50% of available points.

Demand-Controlled Ventilation: A ventilation strategy that automatically adjusts outdoor airflow in real time using sensor signals. Most demand-controlled ventilation systems use indoor carbon dioxide levels as an indicator of occupancy—when carbon dioxide levels rise, the system increases ventilation, and when carbon dioxide levels drop, it reduces ventilation.

Demand Response (DR): An intentional reduction in the electricity usage in response to a demand response (DR) event or changes in the price of electricity.

Demand Response Event (Curtailment Event): The period that the utility company asks for a reduction in electricity usage from its program participants.

Dematerialization: The practice of intentionally reducing the total amount of materials used in a building without compromising its function or performance.

Development Density: A measure of the total building floor area or total number of dwelling units on a unit of land relative to the total buildable area on that land.

Development Footprint: Named for the sum of all the areas that are affected by the project's activity in the project site.

District Energy System (DES): A central energy conversion plant that provides thermal energy, shared by a group of buildings.

Diverse Uses: Publicly available businesses that provide daily need goods or services. According to USGBC, diverse uses do not include ATMs, vending machines, and touch screens.

Diversion Rate: Percentage of waste materials diverted from landfill.

Drip Irrigation Systems: Microirrigation systems that drip water to the roots of plants to minimize the use of irrigation water and fertilizers. They are the most water-efficient systems and have very short payback periods.

Dry Ponds (Detention Ponds): Excavated basins that temporarily hold and slow down stormwater during rain events, but remain dry most of the time. By detaining runoff, they reduce downstream flooding and allow water to infiltrate gradually into the ground as the pond drains.

EDUCATION @USGBC: Education portal of USGBC.

Electric Induction Cooking: A cooking method that uses a magnetic field to directly heat pots and pans, eliminating emissions from on-site combustion.

Electrification: The process of replacing non-electric energy sources with electricity.

Embodied Carbon: Measures the total greenhouse gas emissions associated with a product's entire life cycle—from raw material extraction and manufacturing to transportation, installation, and disposal.

Embodied Energy: The total energy consumed resulting from a product's manufacturing, transportation, installation, use, and disposal.

Emergent Properties: Emergence of certain properties in the systems as a result of interaction of individual elements.

Emissivity (Infrared or Thermal Emittance): A measure that shows how much heat or infrared radiation a material can shed back into the atmosphere.

Energy-Recovery Ventilation: A system in which exhaust air is used to condition incoming fresh air. During cold weather, heat from outgoing air is transferred to incoming air, while in hot weather, the process works in reverse to reduce the cooling load.

Energy Use Intensity (EUI): A measurement unit that describes the building's energy usage relative to its size.

Environmental Justice: The principle that every person deserves equal protection from environmental and health hazards, equal access to decision-making, and equitable access to a healthy and sustainable environment.

Environmental Product Declaration (EPD): Disclosure that reports a product's environmental impacts based on life-cycle assessment data. An EPD describes the environmental effects of raw material extraction, energy use, chemical composition, waste generation, and emissions to water, air, and soil throughout the product's sourcing and manufacturing process.

Environmental Tobacco Smoke (ETS): Smoke from another person's cigarette, cigar, pipe, or electronic smoking device. Breathing in ETS is known as passive smoking, secondhand smoking, or involuntary smoking.

Erosion and Sedimentation Control (ESC) Plan: A plan developed to prevent erosion, sedimentation, and stormwater pollution to the water bodies, wetlands, and the whole neighborhood.

Evapotranspiration: The return of water to the atmosphere through evaporation from soil surfaces and transpiration from plants.

Exemplary Performance Points: Additional points generally achieved by doubling the credit requirements or reaching the next incremental percentage threshold.

Extended Producer Responsibility (EPR): A product stewardship policy approach that holds companies producing consumer goods responsible for managing their own products and packaging when consumers are finished with them.

Extensive Vegetated Roofs: Types of roofs that do not include a variety of plants and require little maintenance. Their soil layer is thinner compared with the intensive roofs since they are more designed for the smaller-sized vegetation.

Facility Manager: The professional responsible for overseeing the day-to-day operations, maintenance, and management of a building to ensure it functions efficiently and meets the needs of its occupants.

Fenceline Community: A community located immediately adjacent to or in close proximity to a source of pollution, such as an industrial facility, highway, or waste site, and therefore disproportionately exposed to environmental and health hazards.

Floor-to-Area Ratio (FAR): A density measure calculated by dividing a building's total gross floor area by the total lot (site) area. For example, a 20,000 sq ft building on a 10,000 sq ft lot has an FAR of 2.0.

Flush-Out: The process of supplying large volumes of fresh air to the building before or during occupancy to remove contaminated air and establish the desired level of indoor air quality.

Formaldehyde: A hazardous volatile organic compound commonly found in adhesives used to manufacture composite wood products such as plywood and particleboard. It is one of the most critical indoor pollutants to control.

Functional Entry: Any building opening that is open and used by pedestrians during business hours.

Gallons per Flush (gpf): A water-use metric for flush fixtures (toilets and urinals) that shows how many gallons are used per flush.

Gallons per Minute (gpm): A water-use metric for flow fixtures (faucets and showerheads) that shows how many gallons flow per minute.

Glare: Excessive brightness, often from high-angle light, that can cause visual discomfort and reduce visibility.

Global Forest Watch: An online platform that international projects may use to understand tree coverage in the project neighborhood.

Graywater: Untreated household water that did not come into contact with toilet waste. Used water from bathtubs, showers, bathroom washbasins, and water from clothes washers and laundry tubs can be examples of graywater and may be used as a flush water in toilets or urinals. This definition can change depending on the local codes.

Green Building: The practice of creating structures and using processes that are environmentally responsible and resource-efficient throughout a building's life cycle, from siting to design, construction, operation, maintenance, renovation, and deconstruction. This practice expands and complements the classical building design concerns of economy, utility, durability, and comfort.

Green Classroom Professional (GCP) **Certificate**: Designed for educators and school facility professionals who want to gain a foundational understanding of green building principles in the context of a school setting, helping them identify what supports or impedes healthy, resource-efficient, and environmentally sustainable learning spaces.

Green Cleaning: The use of environmentally friendly products and cleaning practices to maintain buildings while minimizing harm to human health and the environment.

Green-e: The leading certification program for renewable energy generation in the United States.

Green Infrastructure: Infrastructure to direct the rainwater collected from the impervious surfaces to the vegetation and soil surfaces without routing them to the storm sewer system.

Greenfield: Land that has not been previously developed.

Greenwashing: Refers to the presentation of a product or a material as being more environmentally friendly than it actually is.

Grid-Interactive Building: A building that can adjust how and when it uses electricity depending on grid conditions, helping reduce pressure during periods of high demand, support renewable energy use, and contribute to decarbonization and resilience.

Halons: Chemicals used in fire suppression systems.

Hard Cost: Costs that directly pay for the physical construction of the building, including labor, construction materials, and equipment.

Hazard Assessment: An evaluation of which threats are most relevant to a project by analyzing climate projections alongside the specific conditions of the site.

Health Product Declaration (HPD): Disclosure that provides a product's material ingredients, list of potential chemicals, related concerns, and additional health information.

Heat Island Effect: The phenomenon in which dark-colored, nonreflective surfaces absorb heat during hot weather and release it into the atmosphere, raising local temperatures.

Heat Pumps: An energy-efficient system for heating and cooling spaces and for heating water, draws thermal energy out of water, air, or the ground into a refrigerant and then transfers that heat to another place.

Impervious Surface: A surface that contains less than 50% perviousness.

Inclusive Design (Universal Design): A design approach that aims to make buildings accessible to people of all ages, abilities, and demographics.

Indoor Air Quality Management Plan: A plan developed to protect the indoor air quality for construction workers and building occupants.

Infill Sites (Infill Developments): Sites in which at least 75% of the total area was either previously developed or already being used for other purposes in urban areas.

Inherently Nonemitting Materials: Materials with very low or no VOC content.

Integrated Pest Management (IPM): A sustainable approach that combines knowledge about pests, nature, pest prevention, and control methods that minimize pest infestation and damage while minimizing hazards to the building occupants, the property itself, and the environment.

Integrated Process: An approach that emphasizes early and ongoing coordination among the project team. Owners, architects, and engineers (structural, electrical, mechanical), along with facility managers and other key parties, work collaboratively starting in the predesign phase to align goals and make better decisions.

Intensive Vegetated Roofs: Green roofs with deeper growing media that can support a wider variety of plants (including larger plants), but typically require more structural support and maintenance than extensive roofs.

Intergovernmental Panel on Climate Change (IPCC): The United Nations body responsible for assessing the science related to climate change.

Invasive Plants: Plants that spread and damage the environment by taking over the adjacent existing native and adapted plants.

Islanding: The capability of an on-site energy system to disconnect from the main grid and operate independently during an outage, allowing the building to safely use its own energy resources and continue providing essential services.

Land Trust: A nonprofit organization that works on conserving lands.

Landscape Water Requirement (LWR): The amount of water that the landscape of the site will require during the site's peak watering month.

LEED AP with Specialty: LEED credential, created for professionals with advanced knowledge in green building practices and specialized in a particular LEED rating system.

LEED Campus Program: A program designed for multiple buildings on a shared site under the control of a single entity. It allows project teams to document shared site-related credits once and apply them across all individual building certifications on the campus.

LEED Combined Review: A type of LEED certification review in which the documentation for all design and construction prerequisites/credits is submitted for review at the end of the construction phase (for LEED BD+C and LEED ID+C rating systems).

LEED Fellow: LEED credential created to designate the most exceptional professionals in the green building industry.

LEED Green Associate: LEED credential created for professionals with a proven, up-to-date understanding of green building principles and practices.

LEED Green Rater: Professional who provides in-field verification to LEED Residential projects.

LEED Impact Categories: The high-level outcomes that LEED projects are designed to support. Every LEED prerequisite and credit aligns with one or more impact categories: Decarbonization, Quality of Life, and Ecological Conservation and Restoration.

LEED Interpretation: A precedent-setting official answer to a formal inquiry about the technical details of a LEED prerequisite, credit, or MPR. Unlike other forms of guidance, LEED interpretations apply to all LEED projects, not just the one that requested it. LEED interpretations are administered by USGBC and published in the Addenda database.

LEED Project Administrator: The primary project contact for GBCI. The LEED Project Administrator is the team member who acts as a project manager, overseeing the LEED project and organizing team members for certain tasks, credits, or prerequisites.

LEED Project Boundary: Portion of the site submitted for LEED certification. Defined by the platted property line of the project, including all land and water within it.

LEED Recertification: Required for LEED O+M projects to maintain their certification every 3 years. Projects certified under other LEED rating systems do not require recertification.

LEED Scorecard: Allows project teams to select the credits they plan to pursue and determine whether the total expected points will be sufficient to achieve the targeted level of LEED certification.

LEED Split Review: A type of LEED certification review in which the design prerequisites/credits are submitted for review during the design phase, and both the additional design prerequisites/credits and all the construction prerequisites/credits are submitted at the end of the construction phase (for LEED BD+C and LEED ID+C rating systems).

LEED Steering Committee (LSC): A standing committee established by the USGBC Board of Directors. The LSC sets the strategic direction for LEED development, ensures consistency among all LEED rating systems, and reviews recommendations from other committees to confirm alignment with the overall LEED framework.

LEED Volume Program: A certification approach for organizations planning to certify a large number of new construction buildings or interior spaces based on a common prototype, streamlining certification by focusing on similarities in building design and construction practices regardless of location.

Leverage Point: The point where any action taken in the system can bring about significant results.

Life-Cycle Approach: Evaluates the entire life cycle of a project, product, or service.

Life-Cycle Assessment (LCA): A method that quantitatively evaluates a product's environmental impacts across its life cycle. A cradle-to-grave, or cradle-to-cradle, approach is used in LCA, and the total energy use and other environmental consequences resulting from the creation of that material are calculated.

Life-Cycle Costing (LCC): A method that estimates a product's total cost over its life, including initial cost, operation, maintenance, service/repairs, and end-of-life disposal.

Light Shelves: Horizontal, light-reflecting overhangs that are positioned to reflect the daylight into the desired area of the building.

Light Tubes: Reflective tubular devices that capture daylight at the roof and channel it into interior spaces to provide natural light deeper inside a building.

Lighting Power Density (Connected Lighting Power): The amount of power used per square foot or square meter when all lighting is turned on. Appropriate levels vary by space type.

Linear Approach: A traditional project delivery approach where each team member works separately in sequence, then hands their work off to the next person, with limited collaboration across disciplines.

Load Shedding: The intentional action by a power utility to reduce the load on the power system in order to prevent a total failure of the system.

Load Shifting: Storing energy during off-peak hours and using it later during peak-demand hours to reduce peak loads and electricity costs.

Location Efficiency: A measure of how well a project's surroundings reduce the need for driving by providing easy access to transit, services, and amenities within walking or cycling distance.

Low-Impact Development (LID): A rainwater management strategy that mimics natural site hydrology by managing stormwater close to where it falls. LID strategies include decreasing impervious surfaces and increasing vegetation on-site. LID is used to slow down the flow of rainwater out of site and ensures that the rainwater does not get contaminated.

Makeup Water: Water used to replace the lost water in open systems.

Masking Systems: Equipment used to reduce the background noise in spaces.

Material Circularity: A system in which resources are continuously reused and recycled, reducing dependence on virgin materials and minimizing waste.

Material Efficiency: Using less material to perform the same function or achieve the same results.

Minimum Efficiency Reporting Value (MERV): A rating system that measures how effectively an air filter removes particles from the air. Air filters with higher MERV values will remove more particles from the air, but more energy will be consumed in passing the air through them.

Minimum Program Requirements (MPRs): The minimum standards/characteristics that every LEED project should meet to pursue LEED certification. MPRs state whether a project can pursue LEED certification or not. They provide guidance on the types of projects that are eligible for LEED certification, protect the integrity of the LEED program, and reduce the number of issues that arise during the certification process.

Monitoring-Based Commissioning (MBCx): The process of using software to monitor real-time data from building automation systems and meters, enabling building operators to detect faulty equipment, identify unusual energy consumption patterns, and address issues as they occur.

Monitoring-Based Commissioning Provider (MBCxP): A contracted third party or qualified in-house staff person responsible for leading the monitoring-based commissioning process.

Mulching: A protective layer applied to the surface of soil that will help to keep the roots of the plants cool and therefore prevent evaporation.

Native Plants (Indigenous Plants): Plants that occur and develop naturally in a specific location.

Natural Refrigerants: Refrigerants that occur in nature's biological and chemical cycles without human involvement, such as carbon dioxide (CO_2), water (H_2O), ammonia (NH_3), air, and hydrocarbons such as propane, ethane, and butane.

Negative Feedback Loops: A feedback loop where a change causes a response in the opposite direction to bring the system back toward balance. If a room gets warmer than the set temperature, the thermostat will send a signal to the air conditioning, and the air conditioning will stop blowing warm air.

Net-Zero Energy Project: A project that produces at least as much renewable energy as it consumes on an annual basis.

Nonpoint Source Pollution: Type of pollution in which its source cannot be identified and which generally results from multiple sources.

Nonpotable Water: Water that does not meet the human consumption standards.

Nonprocess Energy (Regulated Energy): The energy consumed by the items that are used to condition spaces and maintain comfort and amenities for building occupants.

On-Site Water Retention: The practice of collecting and retaining rainwater that falls onto pervious surfaces on the project site.

Open Systems: Systems that constantly consume other items, use them, and produce waste at the end.

Open-Grid Pavement: A pavement system with at least 50% unbound.

Operational Carbon: Emissions produced by running the building on a daily basis, such as energy used for heating, cooling, lighting, and powering equipment.

Overage: Extra material ordered as a precaution during construction that often goes unused. Minimizing overage through careful design and procurement reduces waste.

Owner's Project Requirements (**OPR**): A document that outlines the owner's vision and performance requirements for a project, including the LEED objectives the project needs to accomplish.

Particulate Matter (PM): A mix of tiny solid particles and liquid droplets floating in the air, most of which can only be detected using a microscope. PM can carry contaminants such as heavy metals, mold spores, dust mites, combustion by-products, pest droppings, and chemicals from pesticides, cleaning products, or building materials.

Passive Daylighting: A system that collects sunlight using static, nonmoving elements such as windows, glass doors, skylights, light tubes, and light shelves.

Passive Survivability: A building's ability to keep occupants safe by maintaining critical life-support conditions even when services such as power, heating fuel, or water are lost.

Pilot Credits: Credits being tested across all rating system types and credit categories and include credits proposed for the next version of LEED.

Pilot Alternative Compliance Paths (Pilot ACPs): Compliance paths that USGBC is considering for inclusion in future versions of LEED, which can optionally be used in place of existing LEED credits.

Plug Loads (Receptacle Load): Represents the electrical use by all the equipment that is connected to the electrical system via electrical receptacles.

Positive Feedback Loop: A feedback loop where a change causes more change in the same direction—A produces B, and B produces more of A. For example, in an interest-earning savings account, the balance earns interest, and as the balance grows, it earns even more interest, which accelerates the growth further.

Postconsumer Recycled Content: The recycled content of a used material. For example, recyclable printer paper can be sent to recycling after being used and can become a part of new printer paper. Other types of materials with postconsumer recycled content can be aluminum cans, water bottles, most glass, wood and steel products, newspapers, and more.

Potable Water: Water that is approved for human use that meets or exceeds US Environmental Protection Agency drinking water quality standards (or a local equivalent outside the United States.)

Preconsumer Recycled Content: The content of a material that is recycled before reaching any consumer. An example is sawdust generated during the manufacturing of a wood product that is recycled to be used inside an MDF board (medium-density fiberboard).

Preferred Parking: Parking spaces that are closest to the main entrance of a building.

Prerequisites: The minimum requirements that all buildings under a certain rating system must meet in order to achieve LEED certification.

Previously Developed Site: A site that contains at least 75% previously developed land.

Prime Farmland: Land that is used for producing food, feed, forage, fiber, and oilseed crops or is available for these uses, as determined by the U.S. Department of Agriculture's Natural Resources Conservation Service.

Prius Effect: The idea that people change their behavior when they can see real-time feedback. In other words, people can respond to something only if they have immediate information about it.

Process Water: Water used by mechanical or other types of systems in buildings such as cooling towers or medical equipment in hospitals.

Project Priorities Library: A USGBC resource where new credits can be added as they are developed.

Project-Type Credits: Credits designated by USGBC to address the specific, unique needs of a particular adaptation or building application, such as the Places of Respite credit available only to healthcare projects.

Radon: A radioactive gas naturally found in soils, rocks, and water bodies. Exposure to radon is harmful to human health and is the second leading cause of lung cancer after smoking.

Rainwater Harvesting: An aspect of rainwater management that collects and filters the rainwater to be reused as an alternative to potable water.

Rapidly Renewable Materials: Natural materials that can replenish within 10 years.

Reclaimed Water: Former blackwater that has been treated and purified for reuse.

Reference Soil: The site's original, undisturbed native soil—the soil condition that existed before development and that is used as the baseline for comparison (for example, when evaluating soil restoration or soil health).

Refrigerant: Substances used to transfer heat.

Regenerative Design: A design approach where the project is intended to give back more than it takes—producing more energy than it consumes, generating little to no waste, and actively restoring ecosystems and habitats that development has degraded. Excess energy produced on site can be returned to the utility grid, making the building a contributor to the energy supply rather than just a consumer.

Regional Alternative Compliance Paths (Regional ACPs): Credit requirements that may optionally be applied in specific regions of the world to address local standards and conditions.

Renewable Energy Certificates (RECs): Tradable, intangible instruments representing the environmental and other non-power attributes of renewable electricity generation. One REC generally represents 1 megawatt-hour (MWh) of electricity generated from an eligible renewable resource and delivered to the grid.

Renewable Energy: A type of energy that is derived from renewable sources. Renewable energy includes solar, wind, wave, biomass, and geothermal power, plus certain forms of hydropower.

Resilience: A building's ability to prepare for, withstand, respond to, and recover from disruptive events.

Retrocommissioning: A systematic process for analyzing and optimizing building system performance involving a commissioning provider. It includes testing of building systems and documentation of any measures needed to improve performance.

Reverberation Time: The time span between when a sound is produced and when it dies away.

Scope 1 Energy: Direct energy generated from sources owned or controlled by the organization, such as on-site fuel combustion.

Scope 2 Energy: Indirect energy purchased from external sources, such as electricity, steam, or district heating and cooling.

Scope 3 Energy: Indirect energy from sources not owned or directly controlled by the organization, such as employee commuting, business travel, or supply chain activities.

Sensitive Lands: Ecologically sensitive areas such as prime farmland, floodplain, habitat, water bodies, or wetland.

Site Assessment: A part of the integrative process, which clearly shows the project teams the properties of the site, including its topography, hydrology, climate, soil types, water availability, and human health effects.

Smart Growth: A neighborhood development approach that protects undeveloped lands and contributes to project development in locations near jobs, schools, shops, and other diverse uses.

Social Determinants of Health: The conditions in which people are born, live, and work that play a significant role in shaping well-being, including factors such as access to clean air, safe housing, green spaces, nutritious food, and quality education.

Soft Cost: Costs needed to plan, design, and manage a project that do not physically become part of the building. Soft costs include items such as design and consulting fees, project management and supervision, permits, legal fees, financing, insurance, and taxes.

Solar Reflectance (SR): Shows the solar energy that is reflected by a surface on a scale of 0 to 1. A black surface will have a SR of 0 while a white surface will have a SR of 1.

Solar Reflectance Index (SRI): Indicates a material's ability to stay cool by reflecting solar radiation and emitting thermal radiation. Thus, both the solar reflectance and emissivity of a material will be combined to rank the material.

Source Control: The practice of identifying sources of indoor contaminants and taking measures to prevent them from reaching the indoor environment.

Source Reduction: The exact sizing of materials through prefabrication, modular construction, or similar methods to prevent waste generation on-site. It also covers the use of products with less packaging.

Suburban Sprawl: The expansion of populations away from central urban areas into low-density areas.

Sufficiency: A set of measures and practices that avoid unnecessary demand for energy, materials, land, and water while still delivering human well-being within planetary boundaries.

Sustainable Development: Meeting the needs of the present without compromising the ability of future generations to meet their own needs.

Systems Thinking: Refers to the understanding of each and every system of a building while also understanding their relationships and looking at the project as a whole.

Technical Advisory Groups (TAGs): Committees that advise the LEED Steering Committee and consensus committees on specific technical aspects of LEED, such as materials or indoor air quality.

Thermal Bridging: Areas where conductive materials within the building envelope permit unwanted heat transfer in both cold and hot weather conditions.

Thermal Energy Storage: A system that stores heating or cooling capacity for later use, typically in an insulated tank, rather than storing electricity.

Thermal Mass: The ability of a material to absorb and store energy. High-density materials such as concrete and brick require significant amounts of heat to change temperature.

Tier 1 Renewable Energy: On-site renewable energy generation, including systems located on the project site, on the surrounding campus, or on the site of a social impact project where the system is donated and ownership is transferred to the recipient. Eligible resources include solar electric (photovoltaic), solar thermal, wind, and recovered heat from municipal wastewater.

Tier 2 Renewable Energy: New off-site renewable electricity from generation assets contracted to be operational within two years of occupancy or within five years of their commercial operations date.

Tier 3 Renewable Energy: Off-site renewable electricity or fuels that are Green-e Energy certified or equivalent.

Total Volatile Organic Compounds (TVOC): The combined concentration of many VOCs reported as a single value.

Transportation Demand Management (TDM): A set of rules and strategies designed to reduce single-occupancy vehicle trips, including measures such as public transit incentives, carpool programs, employer commuter benefits, limits on parking supply, and policies that encourage walking, biking, and public transit.

Tree Equity Score: A tool that highlights inequities in tree cover and identifies where trees are needed most in each community, helping address disparities by focusing resources in neighborhoods with the highest need.

Triple Bottom Line (**Three Pillars of Sustainability**): A framework that evaluates a project's success across three dimensions: economic prosperity (profit), social equity and justice (people), and environmental stewardship (planet).

Uplight: Light emitted above the horizontal plane that contributes to sky glow.

Volatile Organic Compounds (VOCs): Chemicals that evaporate at room temperature and can be released from building products such as paints, adhesives, sealants, flooring, and furniture. Some VOCs can cause short-term effects like headaches or long-term health problems including cancer.

Walk Score: A measure of a location's walkability on a scale of 0 to 100, based on walking routes to nearby amenities. A higher score indicates more amenities and services within a short walking distance.

Wet Pond (Retention Pond): Pond designed to hold a specific amount of water indefinitely.

Working Groups: Groups established by the LEED Steering Committee or a consensus committee to complete a specific task with a clearly defined scope and timeline.

Xeriscaping: Type of landscaping that does not need any irrigation.

Zero-Lot Project: A type of project in which the building footprint extends all the way to one or more property (lot) lines, leaving little or no space for setbacks on those sides, except for limited access areas (e.g., entryways, walkways, drive aisles, or service access).

INDEX

I

L

M

N

O

U

V

W

X

Z

REFERENCES

CHAPTER 1

1 J.F. Kenny, N.L. Barber, S.S. Hutson, K.S. Linsey, J.K. Lovelace, & M.A. Maupin. Estimated use of water in the United States in 2005: US Geological Survey Circular 1344, (2009).

2 D.M. Roodman & N. Lenssen "A Building Revolution: How Ecology and Health Concerns Are Transforming Construction," Worldwatch Paper 124 (Worldwatch Institute, 1995).

3 Energy Information Administration, EIA Annual Energy Outlook (EIA, 2008).

4 US Department of Energy Annual Outlook 2008.

5 United Nations (2024, August 29), The sustainable development agenda — sustainable development goals. United Nations. https://www.un.org/ sustainabledevelopment/development-agenda/

6 Turner, C. & Frankel, Energy Performance of LEED for New Construction Buildings (2008), http://www.newbuildings.org/sites/default/files/Energy_Performance_of_LEED-NC_Buildings-Final_3-4-08b.pdf.

7 US Environmental Protection Agency, Report to Congress on Indoor Air Quality, volume 2, EPA/400/1-89/001C (EPA, 1989).

CHAPTER 5

1 US Environmental Protection Agency, http://www2.epa.gov/smartgrowth/about-smart-growth.

2 US Energy Information Administration, Emissions of Greenhouse Gases Report (December 8, 2009), http://www.eia.doe.gov/oiaf/1605/ggrpt/.

3 International Council on Clean Transportation, Passenger Vehicles, (accessed March 22, 2013).

4 EPA, http://www.epa.gov/climatechange/ghgemissions/sources/transportation.html.

CHAPTER 6

1 Yang, X., Y.Hou, and B. Chen, Observed Surface Warming Induced by Urbanization in East China, J. Geophys. Res., 116(2011), D14113, doi:10.1029/2010JD015452.

CHAPTER 8

1 "Energy Consumption Characteristics of Commercial Buildings HVAC Systems Volume III: Energy Savings Potential" – US Department of Energy.

CHAPTER 9

1 US Environmental Protection Agency, http://epa.gov/osw/conserve/rrr/imr/cdm/pubs/cd-meas.pdf.

2 US Environmental Protection Agency, http://epa.gov/osw/nonhaz/municipal/pubs/msw2009rpt.pdf.

HOW TO ACCESS THE EXAM SIMULATOR

To access the exam simulator, including the end-of-chapter practice exams and the full-length practice exam, simply scan the QR code below and complete the form to create your password. Once your password has been created, you can log in to your account at any time by scanning the QR codes provided at the end of each chapter and in the final practice exam section, then entering your email address and password.

You may also access your account directly by visiting www.projectific.com/account and entering your email address and password. If you need any assistance, please email us at customercare@projectific.com.

www.ingramcontent.com/pod-product-compliance
Lightning Source LLC
Chambersburg PA
CBHW061204240926
51846CB00006B/1106
* 9 7 8 1 9 7 2 6 3 1 0 5 8 *